LOCAL LIVES

Studies in Migration and Diaspora

Series Editor:
Anne J. Kershen, Queen Mary University of London, UK

Studies in Migration and Diaspora is a series designed to showcase the interdisciplinary and multidisciplinary nature of research in this important field. Volumes in the series cover local, national and global issues and engage with both historical and contemporary events. The books will appeal to scholars, students and all those engaged in the study of migration and diaspora. Amongst the topics covered are minority ethnic relations, transnational movements and the cultural, social and political implications of moving from 'over there', to 'over here'.

Also in the series:

Expatriate Identities in Postcolonial Organizations
Working Whiteness
Pauline Leonard
ISBN 978-0-7546-7365-1

Multifaceted Identity of Interethnic Young People
Chameleon Identities
Sultana Choudhry
ISBN 978-0-7546-7860-1

The Invisible Empire
White Discourse, Tolerance and Belonging
Georgie Wemyss
ISBN 978-0-7546-7347-7

Lifestyle Migration
Expectations, Aspirations and Experiences
Edited by Michaela Benson and Karen O'Reilly
ISBN 978-0-7546-7567-9

International Migration and Rural Areas
Cross-National Comparative Perspectives
Edited by Birgit Jentsch and Myriam Simard
ISBN 978-0-7546-7484-9

Local Lives

Migration and the Politics of Place

Edited by

BRIGITTE BÖNISCH-BREDNICH and CATHERINE TRUNDLE

Victoria University of Wellington, New Zealand

Routledge

Taylor & Francis Group

LONDON AND NEW YORK

First published 2010 by Ashgate Publishing

Published 2016 by Routledge
2 Park Square, Milton Park, Abingdon, Oxfordshire OX14 4RN
711 Third Avenue, New York, NY 10017, USA

First issued in paperback 2016

Routledge is an imprint of the Taylor & Francis Group, an informa business

British Library Cataloguing in Publication Data
Local lives : migration and the politics of place. --
(Studies in migration and diaspora)
1. Immigrants--Cultural assimilation. 2. Human territoriality. 3. Ethnicity.
I. Series II. Bönisch-Brednich, Brigitte. III. Trundle, Catherine.
304.2'3'089-dc22

Library of Congress Cataloging-in-Publication Data
Local lives : migration and the politics of place / [edited] by Brigitte Bönisch-Brednich and Catherine Trundle.
p. cm. -- (Studies in migration and diaspora)
Includes bibliographical references and index.
ISBN 978-1-4094-0103-2 (hardback) 1. Immigrants--Cultural assimilation.
2. Emigration and immigration--Social aspects. 3. Space perception--Social aspects. I. Bönisch-Brednich, Brigitte. II. Trundle, Catherine. III. Title. IV. Series.

JV6225.L63 2010
304.8--dc22

2010028777

ISBN 13: 978-1-138-25096-3 (pbk)
ISBN 13: 978-1-4094-0103-2 (hbk)

Contents

Notes on Contributors *vii*
Series Editor's Preface *xi*
Acknowledgements *xiii*

Introduction: Local Migrants and the Politics of Being in Place 1
Brigitte Bönisch-Brednich and Catherine Trundle

PART I: MIGRANTS AND THE POLITICS OF LAND OWNERSHIP

1 Migrant Routes and Local Roots: Negotiating Property in Dhërmi/Drimades of Southern Albania 17
Nataša Gregorič Bon

2 Against the Gated Community: Contesting the 'Ugly American Dream' through Rural New Zealand Dreams 31
Catherine Trundle

3 Past Imperfect: Displacing Hawaiians as Hosts in a 'Drop Out' Community in Hawai'i 49
Lucy Pickering

PART II: LANDSCAPES OF BELONGING

4 'We are Not Expats; We are Not Migrants; We are Sauliacoise': Laying Claim to Belonging in Rural France 67
Michaela Benson

5 Ambiguous Foreigners: Neighbours Share More Than Geographical Space 85
Jacqueline Waldren

PART III: HOUSES AND HOMES: INTIMATE MIGRANT PLACE

6 A Reluctant Locality: The Politics of Place and Progress in Santo Domingo 101
Erin B. Taylor

7 Little Anglo-India: Making Australia 'Local' at St Joseph's Hostel 119
Robyn Andrews

PART IV: CONTESTING URBAN PLACE

8 Invoking a Community of Engagement: Mobility and Place in a Small English Town 135
Karen O'Reilly

9 Negotiating Religious Expression and Citizenry Belonging: Bosnian Experiences in Suburban Melbourne 151
Lejla Voloder

10 Migrants on Campus: Becoming a Local Foreign Academic 167
Brigitte Bönisch-Brednich

Epilogue: The Cosmopolitan Justice of a Direction Home 183
Nigel Rapport

Index *195*

Notes on Contributors

Robyn Andrews is lecturer in the Social Anthropology Programme at Massey University, New Zealand. Her research on Anglo-Indians began with her PhD (an ethnography of Calcutta's Anglo-Indians), and she has expanded her focus to include the Anglo-Indian diaspora, and Anglo-Indian ageing. She has published numerous articles and book chapters, including, 'Living and Working in Calcutta: Jane's Story' in *Working Women: Stories of Strife, Struggles and Survival* (Sage), and 'Christianity as an Indian Religion: The Anglo-Indian Experience' in the *Journal of Contemporary Religion.*

Michaela Benson has a background in sociology and social anthropology. She has previously co-edited a volume for the Centre for Sociology, Anthropology and Politics (CSAP), entitled *Engagements with Teaching and Learning: A View from Anthropology*. She has also co-edited a special issue on gift exchange for *Anthropology in Action* (2008) and a volume entitled, *Lifestyle Migration: Expectations, Aspirations and Experiences* (2009 Ashgate). She conducted 12 months of ethnographic research among British migrants in southwest France (the first ethnography on this topic), on which she based her PhD thesis and a forthcoming monograph entitled, *The British in Rural France: Lifestyle Choice, Migration, and the Quest for a Better Way of Life* (Manchester University Press). In 2008–2009 she was the Sociological Review Fellow.

Brigitte Bönisch-Brednich (editor) is professor of Anthropology at Victoria University of Wellington, New Zealand. She has published widely on issues around affluent migration, oral history, bi-culturalism and popular culture. Her ethnography of German migration to New Zealand *Keeping a Low Profile: An Oral History of German Immigration to New Zealand* has become a core text in a number of undergraduate and graduate programmes on migration in Europe. She is currently writing a book on academic migrants in the global tertiary education sector.

Nataša Gregorič Bon received her PhD in social anthropology from the University of Nova Gorica, Slovenia. Currently she is a Research Fellow at the Institute of Anthropological and Spatial Studies, Scientific Research Centre of the Slovenian Academy of Sciences and Arts. Her research interests focus on the notions of space and place, borders and boundaries, new regionalisms and Europeanisation. Her regional specialisation is in southern Albania where she has so far conducted 14 months of fieldwork. She has several articles in Slovenian and English on such topics as space, pilgrimage and storytelling in southern Albania, including an article in *Tourism, Culture & Communication* on post-communism, locality and belonging.

Karen O'Reilly, Reader in Sociology, Loughborough University, UK, has a background in sociology and social anthropology and a keen interest in ethnographic methods. She is author of *The British on the Costa del Sol* and *Ethnographic Methods* (both with Routledge), *Key Concepts in Ethnography* (Sage), and numerous journal articles in the fields of tourism and migration. Her research in Spain has spanned nearly 15 years and included long-term ethnographic fieldwork, qualitative interviews with groups and individuals, as well as survey methods. Her research interests are in contemporary migrations (and mobilities) and their implications for sociological 'problems' of nation, ethnicity, class, gender, age, community, home and belonging.

Lucy Pickering is an anthropology lecturer in the School of Social and Political Sciences at the University of Glasgow, UK. She gained her PhD from the University of Manchester on embodiment and identity among hippies and 'drop outs' in Hawai'i. Her research interests include drug cultures and liminal spaces, migration and embodiment and recent publications focus on embodiment, countercultural identity and everyday drug use and recovery in Hawai'i and the UK.

Nigel Rapport is Chair of Anthropological and Philosophical Studies at University of St Andrews, UK. He has also been a Canada Research Professor in Globalization, Citizenship and Justice, at Concordia University of Montreal, and Visiting Professor at Melbourne University and Copenhagen University. His books include *'I am Dynamite': An Alternative Anthropology of Power* (Routledge 2003), *Transcendent Individual: Towards a Literary and Liberal Anthropology* (Routledge 1997) and *The Trouble with Community: Anthropological Reflections on Movement, Identity and Collectivity* (with Vered Amit, Pluto 2002). He has edited many volumes, including *Migrants of Identity: Perceptions of 'Home' in a World of Movement* [with A. Dawson] (Berg 1998) and *British Subjects: An Anthropology of Britain* (Berg 2002). His research interests include social theory, identity, individuality, narrative, community studies, representation and globalisation.

Erin B. Taylor is a lecturer in the Department of Anthropology at the University of Sydney, Australia. She conducts fieldwork in a squatter settlement in Santo Domingo, Dominican Republic, and is particularly interested in the localising effects of globalisation, production and consumption and the politics of place. Her recent publications focus on representations of crime and poverty in the media and public opinion, social imaginaries of rural and urban spaces, and classifications of race and class.

Catherine Trundle (editor) is lecturer in Anthropology at Victoria University of Wellington, New Zealand. She has a PhD in Social Anthropology from the University of Cambridge. Her research explored English-speaking migrant groups in Florence, Italy, with a particular focus on charity. She spent 15 months living in Florence participating in expatriate women's groups and clubs and carrying out life history interviews. She has also conducted research on American migration to rural New Zealand. She has a recent chapter in *Lifestyle Migration: Expectations, Aspirations and Experiences* (2009).

Lejla Voloder is a post-doctoral fellow in Anthropology at the School of Political and Social Inquiry, Monash University, Australia. Her doctoral research investigated Bosnian migrant experiences of citizenship in Australia with a particular focus on the ways in which belonging is claimed within multicultural spaces. Her research interests include migrant integration and social inclusion, multiculturalism, and secular and religious belonging. She has recently published an article entitled, 'Autoethnographic Challenges: Confronting Self, Field and Home' in *The Australian Journal of Anthropology*.

Jacqueline Waldren is a fellow at the International Gender Studies Centre at the University of Oxford, UK, and a research associate at the Institute for Social and Cultural Anthropology at the University of Oxford. She is also the Series Editor for Berghahn Books (New Directions in Anthropology). Her current research focuses on gender, immigration and identity, and her publications include *Insiders and Outsiders: Paradise and Reality in Mallorca* (Berghahn 1996), *Tourists and Tourism* (co-ed, Berg 1997), *Anthropological Perspectives on Local Development* (co-ed, Routledge 1998) and many chapters in edited volumes.

Series Editor's Preface

The binary of migration and place can be interpreted as contradictory or complementary: the contrasting of rootlessness and rootedness or the pairing of the push and pulls that motivate and propel the migrant experience. Whilst the inference of the noun migration is movement, that movement is bounded by specific and grounded locations. It is the role of place in the migrant experience that forms the spine of this edited volume. And, though the geographic spread of the places under the spotlight span the globe – from the Caribbean to Australasia, from Albania to rural France and from the East Midlands in England to the island of Majorca – it is the local with which this book is concerned.

Presenting a cast which includes academics, 'rich Americans' and urban squatters, the contributors to this volume have investigated the processes by which their players have negotiated their way through the politics of locality and belonging; the latter being central to the (re)establishment of a settled migrant life in a new place. As the reader becomes aware, in this process, unexpected bedfellows are thrown together and traditional antipathies and practices are cast aside. Ethnic and racial differences become irrelevant in the struggle to restore a sense of community in what has become a dormitory town in England. Religious observance is reconfigured in the context of contemporary political tensions and debates in Melbourne and central European traditions have had to be replaced by those which reflect the influence of the Anglo-Saxon world on an academic institution in Auckland. In some instances, such as in the case of elderly Anglo-Indians in Australia, the sense of, and need for, belonging was linked to the past. Spatial boundaries became irrelevant as the inhabitants of an old age hostel in Melbourne created their own 'local' – one which mirrored former life in Calcutta – in a strategy which exemplifies the way in which the 'glocal' can become a 'contemporary lived reality'.

This is a book which, through the eyes of a number of its contributors, adopts a subjective approach to the experience of migration and which, in certain instances, as the authors become the performers in their discourse, takes up the insider/outsider debate.

Finally, it is significant that all the contributors to this volume come from an anthropological background. And it is even more significant that the theories, themes and discourses presented only serve to underline the increasing interdisciplinarity of migration studies and the importance of introducing cutting edge volumes such as this to the broadest audience.

Anne J. Kershen
Queen Mary University of London

Acknowledgements

This book has evolved from a panel on migration and the micro-politics of place at the 2008 Auckland conference of the Associations of Anthropologists. The overall conference theme was 'ownership and appropriation' and this seemed to be the right place to take up a discussion of emerging ethnographies on the topic of place, movement and politics. The response to our call for papers was enthusiastic and the quality of papers and discussion impressive. We would like to thank Ghassan Hage for his impromptu, thoughtful and inspiring summing up of the panel contributions, and also all of the colleagues who contributed to our panel, but, for various reasons, could not be included in this volume. It has been a truly rewarding experience to work with everyone involved in this project. We are especially grateful to Nigel Rapport who undertook the task of writing the book's epilogue and looked to the future of migration, migration studies and the political-utopian dreams necessary to attempt to solve the many pressing problems of the global subject. Our research assistants, Jemima Sherpa and Carla Rey, supported us during a long New Zealand summer spent editing. They are both fantastic young scholars and we are sure they will excel in their careers. We are grateful to Victoria University of Wellington, for providing research-assistance funding, and the Association of Social Anthropologists (UK), for providing financial support to attend the Auckland conference. We also thank our editor, Neil Jordan from Ashgate, who offered wonderful support in bringing the manuscript to completion.

Introduction: Local Migrants and the Politics of Being in Place

Brigitte Bönisch-Brednich and Catherine Trundle

Movement is only possible in reference to place. Boundaries can only be crossed if boundaries exist. Mobility ontologically implies its twin, stasis. *Local Lives: Migration and the Politics of Place* goes against the current anthropological grain in a field dominated by transnational studies, by refocusing attention toward emplaced belonging. This volume thus explores the micropolitics and everyday rituals of place-making and shows how, in a fluid world of movement, place remains a deeply contested and symbolically rich site in which to constitute the self, even for those on the move. Despite burgeoning research on transmigrant practices, most people still become mobile with the intention of settling once more and of making the new locality a meaningful site for daily life. This volume focuses on migrant relationships with places and place-attachment and also examines what occurs when mobile migrants settle among groups of people who invoke notions of 'localness' and engage in contested politics of belonging. Our focus is, as Nigel Rapport points out, 'on localisation as cultural technique: here is how migrants utilise and engage diversely with ideas of locality' (this volume: 184).

Examining migration and the politics of place allows us to investigate the tensions between permanence and impermanence, movement and settlement. Over the last decade, the paradigm that confines migration exclusively to movement, to transnational lifestyles and transnational identities, has been gradually but decisively challenged. Scholars of migration are realising that place matters, and the social sciences are once again ready to focus on localities as sites within which the social fabric of migrant lives are woven. Migrants create homes, and in the process they transform place and engage in the small politics of daily life.

Transnationalism, Globalisation and Localisation

For the last two decades, scholars of migration have focused on the emergence of transnational and mobile migrant identities. Labels such as hybridisation, diaspora, creolisation and transmigration abound (for example, Hannerz 1987 and 1996, Appadurai 1991, Glick Schiller, Basch and Szanton Blanc 1995, Gustafson 2002 and 2008, Levitt, DeWind and Vertovec 2003). Such studies have emphasised the deterritorialisation of identity, and a 'multidimensional global space with unbounded, often discontinuous and interpenetrating sub-spaces' (Kearney 1995: 549). Transmigrant identities are seen to contain a range of shifting and contextual affiliations to place due to regular cross-border mobility. These accounts stress the dynamic flow of ideas, political and social networks, goods, remittances, 'tradition', capital and investments that destabilise the very boundedness of host and home spaces (Marcus 1995, Massey 1995, Cliggett 2003, Levitt, DeWind and Vertovec 2003). Scholars began to celebrate the emergence of what were termed 'cosmopolitanism' and 'diasporic identity' (Hannerz 1996, Brennan 1997, Bhabha 1998, Vertovec and Cohen 2002). Locality, as Arjun Appadurai argues, 'has lost its ontological mooring' (1995: 204). Scholars of migration have thus recently focused on the creative reappropriation of place, on agency and ongoing liminality, and this has led to the 'decoupling of locality from territory' (Kennedy and Roudometof 2002: 2). Locality has became 'relational and contextual' (Appadurai 1995: 204), constituted through discourse, narrative and imagination as a nostalgic and fictive anchor in a sea of movement (for example, Holy 1998). In the new transnational reality, such scholars tell us, 'the local' reflects a bricolage of meanings and affective investments that underscore migrants' complex loyalties. 'Home' is thus 'neither here nor there … rather itself a hybrid, it is both here and there – an amalgam, a pastiche, a performance' (Bammer 1992: ix). In all such accounts, movement and mobility have become not just the definitive feature of migrant subjectivities but also the analytic key to understanding the modern world (Rapport and Dawson 1998).

Outside migration research, studies of globalisation have also attended to redefining the local, often as a residual or reactionary category of globalisation. The local is cast in a number of ways: as a place within which the global touches down, dispersing or irreparably altering local cultures; as the site of creative and dogged resistance to the global; or as the place within which all those who cannot attain mobility remain trapped. For example, Roland Robertson introduced the Japanese business term 'glocal' into the social sciences, showing how what we now consider to be local is in fact constituted at the 'trans/supra local' level (1995). Harry Redner argues that 'local cultures maintain difference and promote diversity, whereas global culture breaks down difference and tends towards uniformity and homogeneity.' (2004: 47) Zygmunt Bauman sees the world as in the process of polarising into two unequal groups, one of which is increasingly mobile, privileged and global, the other composed of the vast masses which are rendered immobile, underprivileged and forced to lead local lives (1998, see also Castells 1996). In such accounts, stasis and boundedness are portrayed as involuntary, while mobility is cast as a privilege and a social good. The local and the global are also often portrayed as uncomfortably linked within fractured lives. People must retain 'a foothold in the local while juggling with global forces and connections' (Kennedy and Roudometof 2002: 16). This volume demonstrates how people do not themselves make simple distinctions between local and global aspects of their lives; rather, the distinctions between various flows and spaces are subtle, contradictory and shifting. Groups often deem forces and flows from 'outside' as originating from someone else's local rather than an amorphous global ecumene (such as, for example, ideas of 'American imperialism' in Trundle's chapter, this volume).

Some theoretical accounts acknowledge the mutually constitutive nature of globalisation and localisation (for example Robertson 1995, Smith 2001, Beck 2002), and the complex relationship between geographic mobility and attachment to place. As Sara Ahmed et al. argue, 'being grounded is not necessarily about being fixed; being mobile is not necessarily about being detached' (2003: 1). Yet, despite such sophistication, by deploying the dual terms 'local' and 'global' the frame of analysis remains. While the traffic between the two domains is seen to increase, and the borders of where the local and the global begin and end may be mutable and porous, we are still inevitably left with one domain of social life that is centred on the routine, enduring and face-to-face, and another that spans broader frames of time and space and which is based on imagination, memory and the virtual (for example Appadurai 1995, Hannerz 1996).

The chapters of this volume, by contrast, focus on everyday practices lived out in spaces enabled by material forms, memory, imagination, flows and movement. Little distinction is made theoretically between local and global boundaries or forms in order to allow interlocutors to define their own experiences of flow and closure, closeness and distance, the familiar and the foreign. What emerges ethnographically is that boundaries arise in places where we as theorists often do not locate them, while in other instances our participants express great conceptual ease in connecting intimate practices with wider flows. The authors of this volume do not try to resolve such contradictions but, rather, use them to demonstrate the richness and complexity of migrant belonging.

In the modern world, resistance to immigration frequently relies on symbols and resources declared to be immobile and site-specific. Thus 'local' industry jobs, 'local' council investment and resources, and 'local' heritage and land are often seen to be under threat from immigrants. Migrant responses to such challenges often do not rely on notions of mobility and 'global citizenship', and instead articulate complex reconfigurations of the claims against them. Migrants themselves often express emplaced ownership and entitlement to the host society, and appropriate 'local' and place-specific symbols in order to claim belonging (see, for example, Barrier and Dusenbury 1989, Olwig 2002). Far from seeking to re-essentialise locality, this edited volume uncovers the ways in which migrants utilise and must respond to diverse and specific ideas of 'the local'.

By allowing our interlocutors to define spatial and social boundaries contextually, the authors of this book do not offer a totalising model of 'the local', but an ethnographic exploration of the term. It focuses on those migrants who experience place and its boundaries as powerful and real, despite often active mobility and flow. This is not to say that wider outlooks and identity forms, such as cosmopolitanism, might not be also present in the groups studied. Identity is a multi-faceted and often contradictory process, situationally deployed. In the quest to balance a migration literature focused on movement, imagination and networks across borders, we explore how place is experienced as constraining and producing identity, and how emplacement is a conscious goal and underlying experience of migration, even for those who are mobile. As Birgit Meyer and Peter Geschiere demonstrate (1999: 5), 'fixing' is generated within contemporary global identity politics. Ruth Marshall-Fratani (2006: 12) also notes the current rise of ideas of 'autochthony', 'a strategy for the redefinition, closure, and control of liminal or mobile spaces and categories' (see also Geschiere and Nyamnjoh 2000). We argue, therefore, for more attention to be paid to migrants' own assertions of their status and relationships; while we might define them as transnational, they may self-define as 'locals' or another place-based label (see Benson, this volume).

Instead of mobility and network flows, we have focused on the more material and experiential processes of emplacement. This helps to avoid the trap of constructivism, an assumption often visible in accounts of global imagination and identity. Harri Englund argues that a constructivist approach implies that the processes whereby people create lives precede the experience of living them. Thus, he argues, 'Rather than being a place in which migrants come to be situated, the local appears as an achievement that they carve out of the cultural materials that the fact of their movement provides' (2002: 267). Instead Englund offers the concept of 'emplacement', which he defines as 'a perspective in which the subject is inextricably situated in a historically and existentially specific condition, defined, for brevity, as a "place"' (2002: 267). Thus he argues that:

> The focus on place making as emplacement dissipates the local-global distinction not by returning to analytical models that presume cultural and spatial discreteness, but through enhanced sensitivity to situatedness and presence. This perspective is postglobalist because it builds on earlier insights into flows and circulations in a global space, and it recognizes specific sites and terrains as the conditions of their existence and transformation. (Englund 2002: 268)

This, Englund asserts, eliminates the need for the analytic concept of 'localisation'. We agree, and do not seek to create a model that defines the boundary between where the local ends and the global begins, or indeed how they interact. The chapters of this volume instead illustrate the diverse and subtle interlinking of varying degrees of space, place, imagination, experience and boundary-making processes. 'The local', however, remains significant as a powerful ethnographic and empirical idea in a range of settings today. The groups presented in this volume continue to make distinctions between places, and often defend 'locality' in relation to their experience of living in place.

Community and the Politics of Place

Just as research on migration moved from integration studies to incorporate theories of transnationalism and globalisation, 'community studies' were largely abandoned in order to achieve new approaches to understanding culture, society and place. Community studies, especially in anthropology and European ethnology, was rightly critiqued in the 1980s and 1990s for creating representations of communities as bounded, timeless, culturally consistent and geographically rooted (for example Marcus and Fischer 1986, Rosaldo 1989, Gupta and Ferguson 1997). Such representations have, as James Clifford points out, probably been of more benefit to the anthropologists than to the groups under study (1997: 22–3).

These place-based demarcations and boundaries permitted the construction of a community that the researcher could understand in its totality and present as a holistic case study. Notions of history and change, of connections beyond geographical boundaries, ties and networks, as well as the onset of modernity, were often traded for a stable representation of consistency and essence.[1] Forces 'beyond' the community, such as the state, were often presented as antagonistic threats to the existence of the imagined village, which thus became an enclave apart from modernity.

One powerfully phrased criticism of this approach is found in Henry Glassie's *Passing the Time in Ballymenone* (1982), where he reflects on his own position inside the village. He argues that by understanding the community as 'socially and culturally cohesive, sacred, organic, and unified, rather than ... fragmented into experiential and conceptual alienation, scholars praised the little community – and stripped it of history. All it could do was to wait in a dream state for the diffusion of ideas from without to crack its stasis.' (1982: 148) From the 1980s, after the 'crisis of representation'[2] and the growing interest in symbolic constructions of culture, community studies were rendered largely obsolete in anthropology. Today 'community' is a problematic term not just because it carries uncomfortable baggage from past academic frameworks, but also because, in popular use, it is

1 For summaries of this history see Curtis (2008: 402–5), Amit (2002: 13–20), Frykman et al. (2009: 17–20) and Welz (2009). Note that some anthropologists have argued that this representation of anthropology before the 1980s is simplistic. For example, Bashkow shows how notions of hybridity, diffusion and cultural contact were an intrinsic feature of Boasian anthropology (2004). Furthermore, in the 1950s and 1960s, British anthropologists such as Mitchell (1956) and Epstein (1958) explored new forms of urban community in the African Copperbelt, examining how ethnic and tribal identities were being reshaped within and across new spatial locations. Both examples illustrate that dynamic notions of community, place and culture do pre-date the interpretive turn, although such approaches were in the minority.

2 At this time anthropologists routinely and reflexively began to consider the politics of ethnographic field methods and form of anthropological writing. This was prompted largely by internal critiques, a postmodern turn in the humanities, feminist and post-colonial writing and Said's notion of Orientalism (Denzin and Lincoln 2005: 18).

deployed politically and economically to enact new forms of power (Creed 2004: 57). Current academic uses of the term thus understandably contain a degree of defensive caution.

How then might we utilise ideas of 'community' without resurrecting the romantic, static or governmental meanings of the term? One approach has been to focus on how ideas of community are used instrumentally, particularly in the construction of group identity boundaries (Barth 1969, Cohen 1985, Mitchell 1998). Notions of community, many scholars have demonstrated, are never deployed neutrally, but are rather bound up with forms of empowerment, distinction, governance and resistance (Rose 1996, Baumann 1996). For example, Jennifer Curtis argues that, in loyalist Belfast in the 1970s, 'spatially linked solidarities inspired and limited political agency … for "community," understood as emplaced social relationships, became the motivation, terrain, and subject of struggles with the state' (2008: 400). A focus on instrumental community, however, suffers the same pitfalls as does the idea that locality is 'constructed'. Forms of emplaced relationality are not only used and created by agents, but also experienced on a daily basis in a phenomenological and material sense. Actors experience community not just as a symbolic tool or an identity, but also as a lived reality of social interaction (Amit 2002).

In a recent special issue of *Ethnologia Europaea*, a team of ethnologists and political scientists (Frykman et al. 2009) has taken community studies in a new direction by focusing on collective beliefs, state action and shared daily practice and talk. Their studies are deliberately comparative and qualitative, resisting over-theorisation of what a community is. They do not essentialise cultural characteristics when discussing regionally based networks, carefully linking relationality, place and politics. This approach avoids creating a bounded sense of space by evoking connections between regional communities and the national and transnational political domains and between the state and its people.

In these recent studies, community is not tied nostalgically to the rural or the monocultural. As Ulf Hannerz has argued, 'home is not necessarily a place where cosmopolitanism is in exile. It is natural that in the contemporary world many local settings are increasingly characterized by cultural diversity' (1990: 249). Therefore, in this volume, rural landscapes contain a range of diverse groups negotiating a shared existence, and the four chapters set in urban spaces (Andrews, O'Reilly, Taylor and Voloder), indicate that seeking emplaced belonging and experiencing place-based relationality are not the preserve of migrants in rural areas pursuing an Arcadian dream.

Place, Politics and Belonging

There are several reasons for re-examining the politics of place within migration. First, we must remind ourselves that other disciplines did not abandon place in the way anthropologists and sociologists did when studying migration. Human and critical geography, philosophy and psychology have continued to discuss place (Tuan 1977, Casey 1997b, de Blij 2009). Doreen Massey (1995) claims that places do not have clear boundaries, are constantly changing and are host to often multiple and divergent identities. Arif Dirlik suggests that social scientists should start seeing place, not as a legacy, 'but as a project that is devoted to the creation and construction of new contexts for thinking about politics and production of knowledge' (1999: 40). This suggestion has been taken up in order to analyse the politics of place and how localities in close 'spatial proximity' to each other act as sites of connectedness and interplay (Amin 2004: 38, see also Friedmann 2005).

Within psychology there has also been an increased interest in 'place identity' and the emotional connections people feel to locality (Proshansky, Fabian and Kaminoff 1983, Tuan 1980). From a developmental perspective, the self can only come into being by distinguishing itself, through perceptual modes, from its environment. 'Place identity as a cognitive sub-structure of self-identity consists of an endless variety of cognitions relating to the past, present, and the anticipated physical settings that define and circumscribe the day-to-day existence of the person' (Proshansky, Fabian and Kaminoff 1983: 62). Elena Liotta (2009: 4–5) thus points to the need to include ideas of territorialism in research on the psyche of migrants and the processes of making new homes, with particular emphasis on the sensory aspects of migration (see Andrews, this volume).

The chapters of this volume demonstrate that attachment to place is a significant aspect of modern migration. Place attachment involves 'the interplay of affect and emotion, knowledge and beliefs, and behaviours and actions in reference to place' (Low and Altman 1992: 5, see also Gustafson 2001). In a phenomenological and materialist vein, this book demonstrates that places are enduring (as well as transforming), because they 'gather', as Edward S. Casey argues, 'experiences and histories, even languages and thought' (1997a: 24). For most migrants, even transnational ones, homes must be created. Migration studies have amply demonstrated that such 'homes' can be rooted in diasporic imagining, international networks or collective transnational memory. Yet we argue that 'homes' are also rooted in land, houses, towns, material resources and place-based social networks. 'Homing', as Ahmed et al. show, is as much a material process as a cognitive one; 'affective qualities of home, and the work of memory in their making cannot be divorced from the more concrete materialities of rooms, objects, rituals, borders and forms of transport that are bound up in so many processes of uprooting and regrouping' (2003: 9).

The desire for place attachment means that notions of 'residence' are more central to migrant lives than has been recently acknowledged. Michael Savage, Gaynor Bagnall and Brian Longhurst (2005) show that in Britain today identity is still tied to residence in terms of the life opportunities that residence in a place provides, the distinctions it enables one to forge socially and the affective qualities it engenders. Belonging and being a 'local' thus requires what they term 'elective belonging'. This volume illustrates that their model has wider applicability beyond

Britain. Taking inspiration from Pierre Bourdieu's notion of 'habitus', which leads people to develop feelings of 'comfort' within physical space, they argue that,

> Belonging should be seen neither in existential terms (as primordial attachment to some kind of face-to-face community) nor as discursively constructed, but as a socially constructed, embedded process in which people reflexively judge the suitability of a given site as appropriate given their social trajectory and their position in other fields. (Savage, Bagnall and Longhurst 2005: 12)

Homes *in situ*, then, are seen as sites chosen by migrants for their ability to express and enable identity. To this concept we would add that the materiality and sociality of home places offer opportunities for and restrictions on the ways in which such identities can be reflexively apprehended and enacted, and that the perceived boundedness and 'localness' of specific sites are what commonly draw migrants to a place in order to assert particular identity forms.

Ethnographies of the 'Local Migrant'

A major theme in this volume, and particularly section one, is that of the *unstable boundaries of the local*, of the difficulty in distinguishing between a local person and a foreign person, a guest and a host. These boundaries prove to be porous and contextual, and not simplistically oppositional. Lucy Pickering shows that mainland American migrants to Hawai'i attempted to displace Hawaiian indigenous people as the rightful 'locals', believing instead in a prior mythical people as their hosts. This helped them to legitimise claims to local belonging, and to express their ties to the land. Catherine Trundle illustrates that some American migrants in rural New Zealand attempted to 'Other' compatriot migrants in order to claim a moral right to reside, to own land and to protect the perceived 'local character' of the region. In doing so, they deployed and enabled anti-immigrant discourses used by 'locals.' Correspondingly, Michaela Benson shows that migrants in rural France seeking a particular lifestyle engaged in processes of class-based distinction that downplayed compatriot ties. In Karen O'Reilly's ethnography of an English town, residents did not see migrants per se as threatening the nature of the town but, rather, feared mobile people who failed to develop place attachment and who undermined local regional resource allocations, decision-making processes and a sense of social cohesion. Jacqueline Waldren illustrates the benefits of a historical perspective by showing that, after over 100 years of foreign migration in Deia, Mallorca, guest and host boundaries have become impossible to draw. All these chapters illustrate that movement is not always understood as a liberatory force but can be seen as a threat by 'born and bred' residents and new migrants alike (see, in particular, O'Reilly and Trundle). While the distinctions between locals and non-locals may be hard to draw analytically and ethnographically, the chapters of this volume illustrate that actors, in a range of settings, nonetheless seek to create such boundaries in order to protect or build particular ways of life, practices and identities. Yet they do so creatively, re-working categories often assumed to be fairly stable and commonsensical, such as 'host' and 'guest', in ways that include and exclude through novel means.

An equally important theme is the *materiality of migrant lives*. Nataša Gregorič Bon's chapter illustrates that, for returnee migrants in southern Albania, claiming local ties involved asserting more than just the symbolic and imagined value of land. The returnees made use of spatially configured kin ties, their experiences of working and improving the land over time, and ironically, their right to movement. Robyn Andrew's chapter on Anglo-Indians living in a Melbourne retirement home illustrates that the material, experiential and habitual aspect of living in a shared dwelling were inseparable from imaginings of India and transnational networks and memory work. Thus, sharing Anglo-Indian food and experiencing the materiality of Christian rituals were the processes through which past lives and a present migrant community could be simultaneously evoked. Waldren demonstrates how migrant lives became imprinted on the landscape, inscribed on street signs and headstones and shaped the architectural developments of the town. Foreigners transformed the physicality of the land and, in the process, local understandings of the landscape, just as foreigners, through their incorporation, came to understand the social and moral geographies of the town. For English migrants in rural France, as Benson shows, belonging entailed identifying with a geographic and political region, as well as learning to see and walk the landscape as locals did. In Erin B. Taylor's ethnography of an illegally built settlement in Santo Domingo, migrant residents crafted a precarious sense of belonging through building and maintaining dwellings and shaping streets and public spaces. In the process, they foraged a sense of collective identity and pride.

All the chapters of this volume demonstrate the importance of *place-based identity resources*. Migration studies have recently focused on global imaginaries. They show that, 'people make sense to themselves and others by continually moving amongst a global inventory of ideas and modes of expression' (Rapport 1997: 71). Yet, as Pickering, Benson, Bönisch-Brednich and Trundle demonstrate, even middle class and affluent migrants must contend with fellow residents and their senses of place and actions towards migrants. Migrant imaginations and hopes come up against the constraints of others' imaginations and hopes, and often both make reference to place. This does not mean, however, that locality is a conservative site that can only constrain migrant lives. Rather, place is productive, providing identity resources that migrants often assert and transform in order to craft a sense of emplacement. Voloder, for example, illustrates the complexities of claiming belonging within national and urban spheres. For Bosnian-Muslim migrants in Melbourne, such processes are dialectical. They rely on a nationally deployed rhetoric of inclusion, in the form of Australian secular multiculturalism, which is reflexively critiqued and relied upon to create a Bosnian-Muslim migrant identity. It is through decisions about everyday migrant practices, in private homes and public streets or parks, that an uneasy sense of belonging (and exclusion) can be enacted. Bönisch-Brednich shows that academic migrants operate within an international marketplace and commonly assume that their identities are equally transnational and cosmopolitan. In New Zealand, these migrants often underestimate the cultural particularities of university institutions and the localised practices, unsaid rules and power structures that new migrant academics must learn. They are thus unprepared for their experiences of being 'out of place'. In his epilogue, Rapport offers an evocative and timely suggestion for how we might re-conceive a cosmopolitan anthropology which affirms the rights of all persons to create and break from home spheres.

All of the chapters also testify to the importance of attending ethnographically to the realm of *micro-politics.* Observing 'small politics', as these chapters show, allows us a more intimate understanding of the migration experience and the relationship between place and subjectivity, and to witness migrant strategies and agency in action. In Santo Domingo, Taylor describes how migrant residents claimed locality as part of the politics of daily interaction, talk, habits and routines. They attempted to assert legitimate ownership over their precarious space, civilising and morally revaluing it, while simultaneously locating belonging elsewhere and seeking mobility out of the settlement. For academic migrants in New Zealand, as Bönisch-Brednich shows, settling into their new workplaces required learning culturally appropriate ways of resolving conflict as well as strategies for garnering resources that were idiosyncracies of local institutional structures. Correspondingly, within the contested sections of an Albanian beach, Gregorič Bon's returnee migrants engaged daily in legal disputes and political strategies that required stubbornness and compromises. Following O'Reilly through an English town that is increasingly being abandoned by politicians and some residents, we can see claims for material resources and political voice that would enable forms of placed-based relationality to endure and grow.

This volume argues that the power to defend, define and challenge place is a process in which migrants often willingly and passionately engage. In representing emplaced migrant experience, these chapters demonstrate a continuing human need to find home. Such homes may be created from movement and the crossing of boundaries, or they may imply forms of stasis; commonly they involve both. By allowing migrants, and their meanings and practices, to take centre stage in these chapters, the local has been rendered both conceptually problematic and viscerally real. Yet, in allowing the contradictions of everyday life to permeate our analyses, we see this as an achievement rather than a flaw. In a world of movement, place has not lost its salience as a sphere for making meaning. Rather, it has re-emerged in unexpected realms of social life and is often, paradoxically, tied to new forms of movement.

References

Ahmed, S., Castada, C., Fortier, A. and Sheller, M. 2003. Introduction, in *Uprootings/Regroundings: Questions of Home and Migration*, edited by S. Ahmed et al. Oxford and New York: Berg.

Amin, A. 2004. Regions unbound: towards a new politics of place. *Geografiska Annaler Series B Human Geography*, 86(1), 33–44.

Amit, V. 2002. Reconceptualizing community, in *Realizing Community. Concepts, Social Relationships and Sentiments*, edited by V. Amit. London and New York: Routledge, 1–20.

Appadurai, A. 1991. Global ethnoscapes: notes and queries for a transnational anthropology, in *Recapturing Anthropology: Working in the Present*, edited by R. Fox. Santa Fe: School of American Research Press, 192–210.

Appadurai, A. 1995. The production of locality, in *Counterworks: Managing the Diversity of Knowledge*, edited by R. Fardon. London: Routledge, 208–29.

Bammer, A. 1992. Editorial, in *The Question of 'Home', New Formations* (Vol. 17), edited by A. Bammer. London: Lawrence & Wishart, vii–xi.

Barrier, G. and Dusenbury, V. (eds) 1989. *The Sikh Diaspora: Migration and the Experience Beyond Punjab.* Columbia: South Asia Publications.

Barth, F. 1969. Introduction, in *Ethnic Groups and Boundaries: The Social Organisation of Cultural Difference*, edited by F. Barth. Boston: Little, Brown, 9–38.

Bashkow, I. 2004. A neo-Boasian conception of cultural boundaries. *American Anthropologist*, 106(3), 443–58.

Bauman, Z. 1998. *Globalization: The Human Consequences*. Cambridge: Polity Press.

Baumann, G. 1996. *Contesting Culture: Discourses of Identity in Multi-ethnic London.* Cambridge: Cambridge University Press.

Beck, U. 2002. The cosmopolitan society and its enemies. *Theory, Culture and Society*, 19(1–2), 17–44.

Bhabha, H.K. 1998. Culture's in between, in *Multicultural States: Rethinking Difference and Identity*, edited by D. Bennett. London and New York: Routledge.

de Blij, H. 2009. *The Power of Place. Geography, Destiny and Globalization's Rough Landscape*. Oxford: Oxford University Press.

Brennan, T. 1997. *At Home in the World: Cosmopolitanism Now*. Cambridge, MA: Harvard University Press.

Casey, E.S. 1997a. How to get from space to place in a fairly short stretch of time: phenomenological prolegomena, in *Senses of Place*, edited by S. Feld and K.H. Basso. Santa Fe: School of American Research Press, 13–52.

Casey, E.S. 1997b. *The Fate of Place: A Philosophical History*. Berkeley et al.: University of California Press.

Castells, M. 1996. *The Rise of the Network Society*. Oxford: Blackwells.

Clifford, J. 1997. *Routes. Travel and Translation in the Late Twentieth Century.* Cambridge, MA and London: Harvard University Press.
Cliggett, L. 2003. Gift remitting and alliance building in Zambian modernity: old answers to modern problems. *American Anthropologist*, 105(3), 543–52.
Cohen, A.P. 1985. *The Symbolic Construction of Community*. London: Tavistock.
Creed, G.W. 2004. Constituted through conflict: images of community (and nation) in Bulgarian rural ritual. *American Anthropologist*, 106(1), 56–70.
Curtis, J. 2008. 'Community' and the re-making of 1970s Belfast. *Ethnos*, 73(3), 399–426.
Denzin, N.K. and Lincoln, Y.S. 2005 Introduction: the discipline and practice of qualitative research, in *The Sage Handbook of Qualitative Research*, 3rd Edition, edited by N.K. Denzin and Y.S. Lincoln. Thousand Oaks, CA et al.: Sage, 1–32.
Dirlik, A. 1999. Globalism and the politics of place, in *Globalisation and the Asia-Pacific: Contested Territories*, edited by K. Olds et al. London and New York: Routledge, 39–56.
Englund, H. 2002. Ethnography after globalism: migration and emplacement in Malawi. *American Ethnologist*, 29(2), 261–86.
Epstein, A.L. 1958. *Politics in an Urban African Community*. Manchester: Manchester University Press.
Friedmann, J. 2005. Place-making as a project? Habitus and migration in transnational cities, in *Habitus: A Sense of Place*, edited by J. Hillier and E. Rooksby. Aldershot: Ashgate, 315–33.
Frykman, J., Hammarlin, M., Hansen, K., Rothstein, B., Olofsdotter Stensöta, H. and Schierenbeck, I. 2009. Sense of community: trust, hope and worries in the welfare state. *Ethnologia Europea*, 39(1), 7–46.
Geschiere, P. and Nyamnjoh, F. 2000. Capitalism and autochthony: the seesaw of mobility and belonging. *Public Culture*, 12(2), 423–52.
Glassie, H. 1995 [1982]. *Passing the Time in Ballymenone: Culture and History of an Ulster Community.* Bloomington and Indianapolis: Indiana University Press.
Glick Schiller, N., Basch, L. and Szanton Blanc, C. 1995. From immigrant to transmigrant: theorizing transnational migration. *Anthropological Quarterly*, 68(1), 48–63.
Gupta, A. and Ferguson, J. 1997. *Culture, Power, Place: Explorations in Critical Anthropology*. Durham, NC and London: Duke University Press.
Gustafson, P. 2001. Roots and routes: exploring the relationship between place attachment and mobility. *Environment and Behaviour*, 33, 667–86.
Gustafson, P. 2002. Tourism and seasonal retirement migration. *Annals of Tourism Research*, 29(4), 899–918.
Gustafson, P. 2008. Transnationalism in retirement migration: the case of North European retirees in Spain. *Ethnic and Racial Studies*, 31(3), 451–75.
Hannerz, U. 1987. The world in creolisation. *Africa*, 57(4), 546–59.

Hannerz, U. 1990. Cosmopolitans and locals in world culture, in *Global Culture. Nationalism, Globalization and Modernity*, (Theory, Culture & Society special issue), edited by M. Featherstone. London: Sage, 237–51.

Hannerz, U. 1996. *Transnational Connections: Culture, People, Places*. London: Routledge.

Holy, L. 1998. The metaphor of home in Czech nationalist discourse, in *Migrants of Identity: Perceptions of Home in a World of Movement*, edited by N. Rapport and A. Dawson. Oxford and New York: Berg, 111–38.

Kearney, M. 1995. The local and the global: The anthropology of globalization and transnationalism. *Annual Review of Anthropology*, 24, 547–65.

Kennedy, P. and Roudometof, V. 2002. Transnationalism in a global age, in *Communities Across Borders: New Immigrants and Transnational Cultures*, edited by P. Kennedy and V. Roudometof. London and New York: Routledge, 1–26.

Levitt, P., DeWind, J. and Vertovec, S. 2003. International perspectives on transnational migration: an introduction. *International Migration Review*, 37(3), 565–75.

Liotta, E. 2009. *On Soul and Earth: The Psychic Value of Place*. London and New York: Routledge.

Low, S.M. and Altman, I. 1992. Place attachment: a conceptual inquiry, in *Place Attachment, Human Behaviour and Environment. Advances in Theory and Research*, (Vol. 12), New York and London: Plenum Press, 1–12.

Marcus, G.E. and Fischer, M. 1986. *Anthropology as Cultural Critique: An Experimental Moment in the Human Sciences*. Chicago: University of Chicago Press.

Marcus, G. 1995. Ethnography in/of the world system: the emergence of multi-sited ethnography. *Annual Review of Anthropology*, 24, 95–117.

Marshall-Fratani, R. 2006. The war of "who is who": autochthony, nationalism, and citizenship in the Ivoirian crisis. *African Studies Review*, 49(2), 9–43.

Massey, D. 1991. *Global Sense of Place*. [Online] Available at: http://www.amielandmelburn.org.uk/collections/mt/index_frame.htm [accessed: 21 February 2010].

Massey, D. 1995. Places and their past. *History Workshop Journal*, 39, 182–92.

Meyer, B., and Geschiere, P. 1999. Introduction, in *Globalization and Identity: Dialectics of Flow and Closure*, edited by B. Meyer and P. Geschiere. Oxford: Blackwell, 1–15.

Mitchell, C.J. 1956. *The Kalela Dance*. Rhodes-Livingstone Papers, No. 27. Manchester: Manchester University Press.

Mitchell, J.P. 1998. The nostalgic construction of community: memory and social identity in urban Malta. *Ethnos*, 63(1), 81–101.

Olwig, K.F. 2002. The ethnographic field revisited: towards a study of common and not so common fields of belonging. *Realizing Community: Concepts, Social Relationships and Sentiments*, edited by V. Amit. London and New York: Routledge, 124–45.

Proshansky, H.M., Fabian, A.K. and Kaminoff, R. 1983. Place-identity: physical world socialization of the self. *Journal of Environmental Psychology*, 3, 57–83.
Rapport, N. 1997. *Transcendent Individual: Towards a Literary and Liberal Anthropology.* London and New York: Routledge.
Rapport, N. and Dawson, A. 1998. *Migrants of Identity: Perceptions of Home in a World of Movement.* Oxford and New York: Berg.
Redner, H. 2004. *Conserving Cultures: Technology, Globalization, and the Future of Local Cultures.* Lanham, Md: Rowman & Littlefield.
Robertson, R. 1995. Glocalisation: time-space and homogeneity-heterogeneity, in *Global Modernities*, edited by M. Featherston, S. Lash and R. Robertson. London: Sage, 25–44.
Rosaldo, R. 1989. *Culture & Truth: The Remaking of Social Analysis.* London: Routledge.
Rose, N. 1996. The death of the social? Re-figuring the territory of government, *Economy and Society*, 25(3), 327–56.
Savage, M., Bagnall, G. and Longhurst, B. 2005. *Globalization and Belonging.* London, Thousand Oaks and New Delhi: Sage.
Smith, L. 2001. *Transnational Urbanism: Locating Globalization.* Oxford: Blackwell.
Tuan, Y. 1977. *Space and Place. The Perspective of Experience.* Minneapolis: University of Minnesota Press.
Tuan, Y. 1980. Rootedness versus sense of place. *Landscape*, 24, 3–8.
Vertovec, S. and Cohen, R. 2002. *Conceiving Cosmopolitanism: Theory, Context and Practice.* Oxford: Oxford University Press.
Welz, G. 2009. The strength of ethnography. *Ethnologia Europaea*, 39(1), 49–53.

PART I
Migrants and the Politics of Land Ownership

Chapter 1

Migrant Routes and Local Roots: Negotiating Property in Dhërmi/Drimades of Southern Albania

Nataša Gregorič Bon

In his recent plea for a 'return of the local' in a world that believes it is 'globalizing' (2009: 1) Peter Geschiere explores the meaning of the 'local' or the '*autochthon*'. The term means 'to be born from the soil', a concept that has acquired vital meaning in the present world of increasing migrations and greater mobility. Geschiere explores the meaning of this term in two different contexts of Africa and Europe. In Africa the local struggles for autochthonity are centred on the nation-state, while in Europe they invoke fierce debates over the integration of immigrants. But despite its different modes and historical specificities, the concept of the '*autochthon*' 'celebrates the primacy of being rooted as something self-evident' and enables 'participation in a world shaped by migration' (Geschiere 2009: 38). The link to the land is central to autochthonity as it gives a strong territorial capacity in the globalising world where more people than ever have begun to assert their identities in ways which are deeply rooted in the local.

This chapter will explore the meaning of 'the local' and 'locality' in the southern Albanian coastal village of Dhërmi (the official Albanian name) or Drimades (the local Greek name) of the Himarë/Himara municipality, and will explore its relation to materiality, migrations and belonging. With the acceptance of the Law on Land[1] in 1991 post-communist Albania, many tensions and conflicts emerged in the village space and influenced the re-organisation of social relations, processes of local identifications and the meaning of land and property. This chapter illustrates how villagers who are 'on the move' negotiate, manage and contest their locality, through which they seek to ensure their ownership and property. Based on 14 months of fieldwork in the village of Dhërmi/Drimades between 2004 and 2008, I focus on the returnees who own tourist facilities on the village's coastal plain and the emigrants who continue to return regularly to their natal village. I argue that when expressing their feelings of locality and belonging people continuously reconstruct their past in order to affirm their present, reconstitute and corroborate their ties to land, control their own labour and income, and negotiate

1 After the fall of communism in 1991 the land which used to be part of the collective property and managed by cooperatives became decollectivised.

their sense of mastery. I focus on locality constructions that involve new meanings through which local people and returnees seek to guarantee their property and reinforce their attachment to land. I argue that, in today's shifting economic and political relations, the meaning of locality relates to a group's sense of rootedness in a particular locale as well as to continuous movements and migrations, which I explain with Clifford's concept of 'dwelling-in-travel' (1997: 2).

Dhërmi/Drimades

The collapse of communism and the ensuing liberalisation of Albania were accompanied by economic, social and political instability, causing massive migrations throughout the country. These migrations were especially apparent in Dhërmi/Drimades and other villages of the Himarë/Himara municipality where a great number of people emigrated to Greece after 1991. Many people had also moved out of the village for educational reasons during communism. After 2000, due to decollectivisation of the land, emigrants from Greece began to return to the village. They started re-constructing old houses and building new ones, as well as tourist facilities on the village's coast. From May to September the village is crowded with emigrants who live and work as manual labourers in Greece during the rest of the year, moving to Dhërmi/Drimades in the summer to run these tourist ventures. Return migration and re-possession and re-management of the coastal plain have brought about social differentiations and arguments over the land upon which the returnees and emigrants re-construct their locality and belonging.

The returnees and other villagers, who all believe they 'originate' from the village, often declare themselves to be locals (*horiani*/*vëndasit*)[2] or 'of the place'.[3] This indicates several specific claims about being 'rooted' to the place of their natal origin – either Dhërmi/Drimades or another village of the Himarë/Himara municipality, their language – the local Greek dialect, and their Christian Orthodox religion. Their self-declarations are formed in contrast to newcomers who moved to the village during the period of communism or, more commonly, those who came after its fall for economic reasons. Newcomers often introduce themselves according to the name of the place from which they have moved. The newcomers who came during the communist period generally live in houses purchased from the villagers or built on land given to them according to the Law on Land. Newcomers who moved to the village after the breakdown of communism generally rent old houses owned by the villagers and do not own land.

In contrast to the people living in areas of acknowledged national Greek minority (Gjirokastra, Saranda and Delvina), the bilingual residents (who speak Greek and Albanian) of Himarë/Himara are not considered to be part of this

2 Throughout this chapter the words in Albanian language are written in *italics*, the words in local Greek are written in *italics and underlined.*

3 '*Apo ton topos*/*nga vëndi*'.

minority. Rather, according to Greek politics and mainstream public opinion, they are considered '*omogheneis*' (co-ethnic) Greeks living in Albania. The status of 'co-ethnicity' gives them the right to apply for Special Identity Cards for Aliens of Greek Descent designating them as 'co-ethnics' and allows them unrestricted passage across the Albanian-Greek and other European Union Schengen borders which, due to massive emigration and long visa procedures, are now almost impassable for other Albanian citizens. This status also entitles them to Greek pensions and Greek health and social insurance.

'Dwelling-in-travel'

In his well known book *Routes* (1997), James Clifford writes that in the present world of mobility and migration, dwelling is no longer bound to the 'local ground of the collective life' where travel is a supplement (Clifford 1997: 3). In the twentieth century travelling became normative and thus one can often speak about 'dwelling-in-travel' (1997: 2). Due to considerable erosion of the terrain, lack of land suitable for cultivation, and various economic, social and political changes, people living in Dhërmi/Drimades and the Himarë/Himara area have been moving to and from the area of today's southern Albania and Epirus, Greece for many centuries (Winnifrith 2002). While these movements brought about a multiplicity of connections between people and places, the administrative (Ottoman period) and political (formation of nation-states) divisions caused differentiations and inequalities. During the communist period, when the totalitarian regime of Enver Hoxha forbade any passing of the state borders and, in addition to other restrictions and reforms (such as land reform and nationalisation of property), limited in-country movement (see Gregorič Bon 2008a: 51–2), the movement between southern Albania and Greece also stopped.

After the breakdown of communism and the ensuing massive migrations (King and Vullnetari 2003, Vullnetari 2007), a significant part of the population migrated to Italy and Greece (Mai and Schwandner-Sievers 2003, Vullnetari 2007: 14), and in later years to the United States and elsewhere in Europe. To paraphrase Clifford (1997), movements and migrations became the norm in post-communist Albania. This is especially the case for Dhërmi/Drimades, where recent returnees who came back to their natal village after retiring maintain relations with their children in Greece. This means that almost every year, especially when they go to the doctor or collect the pensions provided by the Agricultural Insurance Organisation in Greece, they visit their children's families and spend several weeks with them. Continuous connections with Athens, where most of the returnees' children and grandchildren live, are maintained by different bus companies in Himarë/Himara, transporting passengers three days a week during the winter and every day during the summer. Besides the bus connection, in the summer there is also a connection by sea on speedboats travelling between Corfu and Himarë/Himara. In addition to physical contact (such as visits), the locals keep their ties with their children through the Greek mobile phone system which, due to the proximity of Greece (the island of Corfu and Albanian-Greek state border), is available along with the Albanian mobile phone system.

Because of continuous movements through the Albanian-Greek border, maintained before and after the communist period, a number of villagers refer to the state border as the 'road' which connects their village to Greece and Europe. In spite of the fact that in practice few of the villagers travel beyond Greece, they often emphasise their ability to travel 'freely' to the countries of 'western' Europe. They often use this privilege to differentiate themselves from other citizens of Albania, whose travels are restricted by visas that must be acquired through long bureaucratic procedures. Their actual movements, together with their ability to cross the Schengen borders, serve as means by which the people of Himarë/Himara re-construct social and spatial boundaries.

Many elderly returnees often recount stories recalling their ancestors' movements to places over the sea and mountains, dating back to the period before communism (Gregorič Bon 2008a: 7–27). While stories about such mobility relate the village, its area and its people to Greece and Italy, 'civilisation', economic development and general wellbeing, stories of movements over the mountains relate the village and its people to Albania, poverty and a lack of 'civilization' (Gregorič Bon 2008a: 15–20). It seems that the locals' continuous movements and travels to and from Greece constitute their sense of 'dwelling', which is conceptually similar to Clifford's notion of 'dwelling-in-travel'. Through the villagers' self-ascriptions as locals 'of the place' or particular 'locale', and through their recounting of past movements and trading relations (Gregorič Bon 2008a: 15–22) and their access to 'free' passage over the Albanian-Greek and EU Schengen borders (Gregorič Bon 2008b: 83–105), 'place' does not appear as a bounded site but rather as an itinerary, or in Clifford's words, 'a series of encounters and translations' (1997: 11).

In contrast to Appadurai (1996), who sees locality as 'primarily relational and contextual' rather than spatial, I argue that spatiality is constitutive and constituting of the locality in Dhërmi/Drimades and vice versa. While locality may be based on the villagers' sense of being 'rooted' to the place, land and property, it is also based on the villagers' continuous movements and migrations. The latter, along with the endless negotiations of locality and property, means that for villagers place and locale are an 'instantaneous configuration of positions' constituted by a system of signs (de Certeau 1988: 117). Because people's claims of belonging relate to a plurality of places more than to one particular place, I define locality as a process of becoming. As this ethnographic account will show, the sense of a particular locale or land, and consequently locality, is continuously shifting as it is negotiated and managed by people pursuing their practical, personal or social goals. These goals are often related to claims for property and land ownership, which have now become the subject of many heated debates.

Property

The complexity of the processes of denationalisation, privatisation and restitution in Albania brought ambiguous understandings of land and property,[4] which differ from the property relations established in the pre-communist period. Before I describe the complexity of decollectivisation and the problems caused by its application, however, let me pause to define some key concepts in this chapter's continuing debate.

Several anthropologists working in eastern Europe (Kideckel 1992, 1995a, 1995b, Hann 1993a, 1993b, 1998, Verdery 1994, 2003, Abrahams 1996, de Waal 1996, 2005, Creed 1998, 1999, Kaneff 1998, 2000, 2009, Meurs 2001, Lampland 2002, Pine 2004 and so on) discuss various responses to decollectivisation and view property and ownership from different perspectives. In this chapter, Hann's (1998) and Verdery's (1998: 161) conceptualisations of property as a bundle of powers rather than rights provides a useful point of entry to the meanings of land and property. I concentrate on the meaning of property as the 'set of social relations, and organisation of power' that come together through social processes (Verdery 2003: 19).

The process of decollectivisation introduced new differences and inequalities among the people of Dhërmi/Drimades and the Himarë/Himara area. These are being shaped through the reconstruction of individual belonging to locality, which is closely intertwined with individual and collective ownership, as well as people's attachments to land. This chapter shows how the process of decollectivisation and restitution of property changed relationships in the village and influenced people's self-representations, their relation to labour and their belonging to 'their' place. I do not seek to define 'property' and 'locality' but to illustrate and explain how porous and flexible their meanings are, and to question what ramifications this has in the local as well as the national and supranational European contexts.

Land/*ktima*/*toka*

After the fifteenth century most of the area that is now Albania submitted to the Ottoman administration. Yet people living in the isolated mountainous areas of northern Albania (Mirdita) and some places in the south (such as Himarë/Himara, Dhërmi/Drimades, Saranda, Delvina, Gjirokastra and parts of Korca) managed to keep their autonomy through local administration, 'customary law' and religion in exchange for paying taxes (Rusha 2001). The people of Dhërmi/Drimades kept their own property, which, besides a house and adjoining small garden, also comprised of agricultural land, pasture and small woodlands. While the house and agricultural land were inherited by partible inheritance (each male heir received

4 Whereas the people of Dhërmi/Drimades generally use the terms 'property' (*pronë*/*idioktisia* or *periousia*), 'proprietor' (*pronar*/*idioktitis*), 'land' (*toka*) or 'estate' (*ktima*) in their daily discourse, I use the term 'property' throughout this chapter.

a lot or a house), the pastureland and small woodlands were inherited by the impartible principle – all the men of one patrigroup[5] (*soi/fis*) inherited a portion of the land together. In theory, there was a strong agnatic preference for dividing the land and property exclusively among the sons. The settlement pattern tended to be patrilocal and marriages, endogamous within the village, were based on pre-marriage agreements between two patrigroups. These decisions were based on the perceived need to keep ownership of the land within the village and to preserve the Christianity of the area.

With the beginning of communism in 1945 came land collectivisation and the foundation of cooperatives. In 1957 Dhërmi/Drimades became part of the agricultural cooperative. Land previously owned by individual proprietors and patrigroups became part of the state cooperatives. After 1980 private houses became part of the cooperatives too. Some of them were used for the purposes of the cooperative while others remained inhabited by their previous owners. The communist party relocated several people to the village from other places in the Albanian state. They lived in the village as public workers, such as teachers, doctors, policemen, etc. Many local youth requested relocation to the capital, Tirana, or the coastal city of Vlora for educational purposes. After completing their studies many found jobs in the cities, where they married and settled with their families. The marriage pattern gradually changed from village endogamy to include exogamous marriage outside of the village.

A year after the fall of communism in 1990 the government passed the Law on Land (Law No. 7501 on Land, 19 July 1991). The law declared that land, which used to be state property and managed by the cooperatives, should be divided equally between the members of the cooperative. Thus, every member who once worked for the cooperative should own a proportionate piece of land, its size based on the size of the entire area of the cooperative's operation. This law resulted in many disputes between the local people of Dhërmi/Drimades, as well as those in other areas that had managed to keep their autonomy during the Ottoman conquest. Internal population movements from rural to urban areas and vice versa between the 1960s and 1970s, instigated by the communist government, had changed the population of the village. After the fall of communism several locals, both those who during communism had worked in the village cooperative and those who had moved to the cities, migrated to Greece. While most of those migrants who had worked in the village cooperative never returned to Dhërmi/Drimades, those who had lived elsewhere in Albania during communism have now begun to return to their natal village. Therefore there are more returnees living in the village than there are people who remained there during communism. As most of the returnees were not members of the village cooperative they were not given the right to use the land. Because of this the Himara Community, the Himarë/Himara municipality and the local people decided together to abrogate the Law on Land and implement a

5 Patrilinear descendants of an apical ancestor as well as other blood and fictive kin assimilated into the line of descent.

consensus arrived at by the population of Himarë/Himara, involving the restitution of the land to its previous proprietors.

In 2004 the government passed another Law on Restitution and Compensation of Property for a few expropriated feudal Albanian families who had owned land in Tirana before communism. As their land now serves the public interest, the government decided to compensate them. Because of financial shortages in the state budget, they compensated these owners with costal land in the Himarë/Himara municipality, which, according to the political elite, is state property. While the law on restitution and compensation has 'silenced' the conflicts and discordances within particular kin groups and within the village, it has also led to conflicts and discordances between the local community and the leading political elite.

Negotiating the 'Local'

Tourism on the Dhërmian/Drimadean coast dates back to the 1960s, when the communist Labour Party transformed some of the old buildings, formerly used as warehouses for storing oranges, olive oil and olives, into a hotel, a government villa and a Workers' Camp. In the first few years after the collapse of communism and the beginning of privatisation, the state buildings and the land that had been owned by the Communist Labour Union were leased to people who had come from other parts of Albania. In the years after 1997 and especially after 2000 when the Himarë/Himara area was acknowledged as a municipality and the national road from Dukati to Palasa was reconstructed, tourist facilities on the coastal plains of Dhërmi/Drimades grew in number. The owners, originating from other parts of Albania, built nine new buildings that are now used as guest houses, room rentals, a bungalow site, hotels and a disco bar. Except for one (the smallest hotel), all are situated on the northern side of the Potami (literally 'stream'), a small stream flowing into the sea. In 2000, local people also began to build tourist facilities, which are primarily located on the southern side of the Potami. They built nine facilities, including guest houses, small hotels, a bungalow site, a restaurant and a night club. The majority of locals who run tourist facilities worked as emigrants in Greece for at least a few years. Three of them still live in Greece for most of the year, returning to the coast of Dhërmi/Drimades only in the summer season. Another three local owners of tourist facilities live in larger towns such as Vlora and Tirana in Albania and are present here only during the summer months. The last three owners live in the village permanently, but often go to Greece where they visit their children, see the doctor and renew the Special Identity Cards for Aliens of Greek Descent, if needed.

Kosta, who moved out of the village when he was 15 years old, returned in 2003 with a plan to build a restaurant on the so-called 'local side' of the coastal plain. He had applied for relocation to Tirana for educational purposes. There he married Ariadne, from Saranda where the Greek minority lives. A year after the fall of communism they and their three children migrated to Greece. When I asked why they had returned to the village, Kosta replied: 'Because I own land here and I mean something. All my life I have been working for other people and now I finally work for myself and for the future of my children.' The majority of returnees in the village described their reasons for returning in similar terms, often concluding that land is 'like their bank'. When Kosta returned to the village, he claimed half of the 2,000 square metres of coastal land on which his cousin Andrea (the son of a brother of Kosta's father) had built a bar in 2001. Like Kosta, Andrea too had moved out of the village for educational purposes when he was a teenager. After finishing school in Vlora, Andrea married a woman from the city. In 1991 he, his wife and children (two sons and a daughter) migrated to Greece. In 2000 Andrea and his wife settled in the village and a year later, with their sons who live in Greece but return regularly to the village in summer months, they built a bar on the coastal plain.

Kosta's claim to a portion of the coastal land led to a dispute between him and Andrea. At first, Andrea resisted giving Kosta any of the land, saying that he had returned to the village first, thus the land should belong to him. As neither of them had worked in the village cooperative, they knew that legally neither had any right to the land. Kosta noted that everybody in the village knew this land belonged to their patrigroup and since he and Andrea are the only heirs currently living in the village, they should split the land in two equal parts. As Andrea's bar was situated in the middle of the parcel he was forced to pull it down after several months of arguing with Kosta. That same year, Andrea built another bar on his part of the land which he and his sons now run. On the other half of the land Kosta and his son Archilea, who returned from Greece and settled in the village in 2004, built a restaurant. The difficulties between Andrea and Kosta continued. Kosta's wife Ariadne often complained about the conflict, which she believed exposed Andrea's carelessness and irresponsibility towards the land. She often recounted the story of the garden behind their restaurant where Andrea had disposed of empty bottles and other rubbish in the years he had had his bar there. She complained that she and Kosta had to put a lot of effort into removing the bottles, disposing of them outside of the village, in order to put in the garden where they now grow most of the vegetables used in their restaurant. She often criticised Andrea's wife as being lazy and lacking good work habits. Gardening is a mode used by Ariadne to establish her physical relationship to the land which, together with her care and responsibility for the restaurant, forms her identity and defines her belonging to the land and the village. Not having been born in the village, she has no ancestral ties to it. When gossiping about Andrea's irresponsibility, she stressed their own conscientiousness regarding the 'proper' management of their restaurant and land. In a slightly reserved manner, Kosta similarly criticised Andrea's irresponsibility and lack of loyalty towards his kin. He often noted that he and Andrea are of the same patrigroup and therefore they should cooperate and help each other. Like Kosta and Ariadne, Andrea and his wife connected their sense of being 'of the place' to locality. They often emphasised their work and effort in building the bar a year after their return from Greece. They also expressed their responsibility and the successful management by their sons, who gained experience managing bars in Greece. Moreover, they pointed out that their bar was busier than Kosta's restaurant.

Besides economic value, land ownership brings cultural capital to returnees and locals, who once worked for the communist government and later, through emigration, for landlords. Verdery (2003: 178) defines this notion as the sense of mastery, which is not only an individual but also a communal matter. In Dhërmi/Drimades the sense of mastery is often related to the meaning of property and locality. For example, when Ariadne declares her labour, care and responsibility for the land, she exposes Andrea's carelessness, irresponsibility, and insufficient management. The sense of mastery is used also by Andrea who boasts about his responsibility and successful management of his bar which is making higher profits than Kosta's restaurant. The sense of mastery is thus continuously shifting and is contextual, relational and sometimes oppositional.

Some months after the local authorities of the Himarë/Himara municipality decided to abrogate the Law on Restitution and Compensation of Property and to implement the Law on Land 'in their own manner', the procedure of land registration began. This procedure was based on verification of the old testaments and other land documentation which date back to the Ottoman administration period and have been preserved either in people's personal archives or in the national archives in the capital, Tirana. Upon submission of these documents the municipality would issue land ownership documents and building permits for tourist facilities on the coast. In 2005, based on one of his grandfather's letters, Kosta registered the 2000 square metres of coastal land upon which he and Andrea had acquired the building permit for the restaurant and bar. The relationship between Kosta and Andrea gradually improved. For example, when Kosta had no electricity because of a delay in his payment, Andrea offered him access to his power line. Since then, Kosta and Andrea have helped each other run their businesses on the coast. 'I prefer to cooperate with troublesome kin rather than let the state to take the land away from me', Kosta said, explaining the sudden change in their relationship. The registration process and enforced returning of ancestral lands strengthened the differences between the local owners, largely returnees, and the owners from other parts of Albania. Thus notions of who owns the land and who does not are flexible and negotiable.

Many owners of tourist facilities who are from other parts of Albania saw their labour as a central component of their propertied identity and considered their purchase of the land from the state authorities to be something that makes them 'real' owners. In contrast, the returnees and local owners of tourist facilities view their land as a source of reconnection with their ancestors. 'This is the land of our grandfathers' is a common explanation of local owners' claims to the land. They often recounted stories about the trading relations that their grandfathers maintained before the communist period with the island and mainland of today's Greece and Italy. Kosta, for example, often recalled a story about the village port, Jaliskari, situated on the north-western side of Potami or on the '*kseni meria*' – the local part – as many of the local owners and other villagers refer to that section of the village's coast. Before the communist collectivisation of private property, the port was owned by a few prosperous patrigroups in the village who anchored their boats there. Some years ago a man from Tirana built a bar there. Kosta and many other local owners often complained that most of the 'foreigner' owners stole land which had belonged to the village patrigroups. The local owners do not consider the 'foreigner' owners to be the 'real' proprietors because, according to their statements, they are not 'of the place' and do not belong to the village space. In contrast, they constitute themselves as being 'of the place'. When claiming this they often express the autonomy of the area, and the trading relations that their ancestors had managed to sustain with what is now Greece and Italy until communism. Moreover, many returnees often speak of their ability to freely cross the Albanian-Greek and other Schengen borders which are hardly passable for the majority of Albanian citizens. The local owners and returnees construct their sense of being cosmopolitan, which they see as a vital part of 'their' local connectedness, on the past movements of their ancestors as well as their own present movements.

Resisting the Law on Land

Following Gupta and Ferguson, who define resistance as a way of shaping the identity of the 'subjects despite its conjectural character' (2001: 19), I argue that resistance in Dhërmi/Drimades is constituted by and constituting of locality. In order to object to the Laws on Land and Restitution and Compensation of Property, the intellectuals of Himarë/Himara (many of whom are living in Greece or the Untied States), along with the association of the Himara Community and the Himarë/Himara municipality, organised Pan-Himarian conferences in 2005 and 2008. At both conferences, local intellectuals, emigrants originating from Himarë/Himara and some returnees discussed the present and future development of the municipality with particular emphasis on tourism and land ownership. The organisers invited the media and in 2008 even invited the Albanian president, Bamir Topi.

But in spite of these events, conflicts between the local people and the political elite continued. In February and August 2008, protests were organised in Himarë/Himara with the local people demanding 'land ownership and autonomy for the area' (as was reported in their own newspaper, *Himarë/Himara*, published in Greek). In 2006, local intellectuals along with the Himara Community set up a website[6] to provide general information about the villages of Himarë/Himara and their history, and also offering a tourist guide and blog discussions on the current property issues.[7] The website is posted in three languages: Albanian, Greek and English. The tourist guide presents the history and tourist attractions of the

6 http://www.himara.eu.
7 http://himarablog.blogspot.com/search/label/Prona.

area, emphasising its autonomy, trading relations, movements to today's Greece before the period of communism, and the Christian religion, using these features to construct the distinctiveness of Himarë/Himara 'region'.

Local efforts to establish an autonomous region transcend the area's geographic borders, yet generate and redefine the meaning of locality. The local intellectuals, municipal authorities and the villagers themselves are striving to find their place in the 'Europe of regions'.[8] Therefore many intellectuals attempt to virtually connect their region to the European Union. This supranational governing body, they believe, could provide them with opportunities to strengthen their regional autonomy and solutions to their land tenure issues.

Conclusion

The process of decollectivisation has influenced changes in the perception of land, which no longer has a collectivised but a market value. As Kosta and many other villagers articulated, ownership of coastal land mainly used for tourism purposes brings economic benefits and represents a long-term investment for many villagers. Like Verdery's (2003: 173–5) example from a Bulgarian village, in Dhërmi/Drimades property also brings responsibility and autonomy to people who once lacked these benefits. Under communism land and labour were held in collective ownership. Now they have become independent and autonomous categories that individual people are able to control.

This ethnographic material illustrates how returnees, through the reconstruction of locality and belonging, reassert their property rights, constitute their identity and reinforce their attachment to place, which they seek to emplace in the 'Europe of regions'. It also exemplifies how returnees, through their sense of rootedness and their mobility, construct their place as a set of encounters and translations. When managing and negotiating their feelings of belonging they expose their past and present movements, and in the process constitute their locality as a form of 'dwelling-in-travel' (Clifford 1997).

This chapter demonstrates how social boundaries between the 'locals' (those 'of the place') and 'foreigners' (those 'out of place') are spatialised and emplaced. In people's daily discourses the coastal plains are divided into a 'local' and 'foreign' side. Social boundaries serve as the means through which returnees constitute their locality and rootedness to place, which is reaffirmed through their present and past

8 The 'Europe of regions' is a specific term used in the European Union which demarcates administrative and transborder Euro-regions. While on the one hand the regions (as they are constituted in the political agendas of the regional policy-makers and economists) are administrative and political, the social or symbolic regions – as Roth (2007: 33) names them – are on the other hand constructed through social differences and nostalgic discourses. These serve peoples' recreation of their regional belonging, which they exclude from the national past, (re)turning to a European present and future.

movements to places in Greece, Italy and elsewhere. The meanings of locality and property are influenced not only by the returnees' and villagers' roots in their natal village, but also by their routes to and from their village. When returnees express belonging through property rights they reconstruct their rootedness to place and expose their ability to move. The meaning of locality is thus ambiguous as it results from the continuous interplay between mobility and rootedness. Returning to Geschiere (2009), in his conceptualisation of the '*autochthon*' or the 'local' he does not point out this interplay between mobility and rootedness, but rather focuses on the peoples' link to the land as the flip side of globalisation. By contrast, the ideas of '*autochthon*' in Dhërmi/Drimades merge mobility and rootedness without contradiction.

I have described some of the ways in which identity and relations among the people were reorganised as the land in southern Albania became decollectivised. The local people's implementation of the Law on Land in their own manner has brought about social conflicts both among kin and among the village inhabitants. It has reasserted the predominance of particular relations over space, privileging the people 'originating' from Dhërmi/Drimades and other villages of Himarë/Himara over those who originate elsewhere, creating social and spatial differences. In such contests, property became a vital subject in the process of constructing and reconstructing relationships, locality and space. Locality implies reference to a place, signifying a sense of rootedness and origins, yet entailing dwelling-in-travel. Property and locality do not have singular references but are context-bounded and contingent. In Dhërmi/Drimades ideals of locality and property celebrate 'rootedness' and gives the returnees a feeling of emplacement in the village space, a space that is shaped by movements and migrations.

References

Abrahams, R. (ed.) 1996. *After Socialism: Land Reform and Rural Social Change in Eastern Europe*. Providence and Oxford: Berghahn Books.

Appadurai, A. 1996. *Modernity at Large: Cultural Dimensions of Globalization*. Minneapolis and London: University of Minnesota Press.

de Certeau, M. 1988 [1984]. *The Practice of Everyday Life.* Berkeley: University of California Press.

Clifford, J. 1997. *Routes: Travel and Translation in the Late Twentieth Century*. Cambridge and Massachusetts: Harvard University Press.

Creed, G. 1998. *Domesticating Revolution: From Socialist Reform to Ambivalent Transition in a Bulgarian Village*. University Park: Pennsylvania State University Press.

Creed, G. 1999. Deconstructing socialism in Bulgaria, in *Uncertain Transition: Ethnographies of Everyday Life in the Postsocialist World*, edited by M. Burawoy and K. Verdery. Boulder: Rowman and Littlefield, 223–44.

Geschiere, P. 2009. *The Perils of Belonging: Autochthony, Citizenship, and Exclusion in Africa and Europe*. Chicago and London: University of Chicago Press.

Gregorič Bon, N. 2008a. Storytelling as a spatial practice in Dhërmi/Drimades of southern Albania. *Anthropological Notebooks*, 14(2), 7–29.

Gregorič Bon, N. 2008b. 'Where are we? In Europe or Albania?' Regionalism as seen by the local people in Dhërmi/Drimades of southern Albania. *Two Homelands*, 27, 83–105.

Gupta, A. and Ferguson J. (eds) 2001 [1997]. *Culture, Power and Place: Explorations in Critical Anthropology*. Durham and London: Duke University Press.

Hann, C. 1993a. From production to property: decollectivization and the family-land relationship in contemporary Hungary. *Man*, 28, 299–320.

Hann, C. 1993b. Property relations in the new eastern Europe: the case of specialist cooperatives in Hungary, in *The Curtain Rises: Rethinking Culture, Ideology and the State in Eastern Europe*, edited by H. DeSoto and D. Anderson. Atlantic Highlands, N.J.: Humanities Press, 99–119.

Hann, C. 1998. Introduction: the embeddedness of property, in *Property Relations: Renewing the Anthropological Tradition*, edited by C. Hann. Cambridge: Cambridge University Press, 1–47.

Kaneff, D. 1998. Private cooperatives and local property relations in rural Bulgaria. *Replika, Special Issue: Central European Hysteria* (English edition), 3, 161–71.

Kaneff, D. 2000. *Property, Work and Local Identity. Max Planck Institute for Social Anthropology Working Papers*, no. 15, Halle, Germany.

Kaneff, D. 2009. Property and transnational neoliberalism: the case of British migration to Bulgaria, in *Accession and Migration: Changing Policy, Society and Culture in an Enlarged Europe*, edited by J. Eade and Y. Vakanova. London: Ashgate, 59–74.

Kideckel, D. 1992. Peasants and authority in the new Romania, in *Romania after Tyranny*, edited by D. Nelson. Boulder: Westview, 69–83.

Kideckel, D. 1995a. *East European Communities: The Struggle for Balance in Turbulent Times*. Boulder: Westview.

Kideckel, D. 1995b. Two incidents on the plains of southern Transylvania: pitfalls of privatization in a Romanian community, in *East European Communities: The Struggle for Balance in Turbulent Times*, edited by D. Kideckel. Boulder: Westview, 47–64.

King, R. and Vullnetari, J. 2003. *Migration and Development in Albania*. Development Research Centre on Migration, Globalisation and Poverty. Sussex: Sussex Centre for Migration Research.

Lampland, M. 2002. The advantages of being collectivized: comparative farm managers in the postsocialist economy, in *Postsocialism: Ideals, Ideologies and Practices in Eurasia*, edited by C. Hann. London and New York: Routledge, 31–56.

Mai, N. and Schwandner-Sievers, S. 2003. Albanian migrations and new transnationalisms. *Journal of Ethnic and Migration Studies*, 29(6), 939–48.
Meurs, M. 2001. *The Evolution of Agrarian Institutions: A Comparative Study of Postsocialist Hungary and Bulgaria*. Ann Arbor: University of Michigan Press.
Pine, F. 2004. Reproducing the house: kinship, inheritance and property relations in highland Poland, in *Distinct Inheritances: Property, Family and Community in a Changing Europe*, edited by H. Grandits and P. Heady. Halle Studies in the Anthropology of Eurasia No. 2. Münster: Lit Verlag, 279–96.
Roth, K. 2007. What's in a region? Southeast European regions between globalization, EU-integration and marginalization. *Ethnologia Balkanica*, 11(1), 17–41.
Rusha, S. 2001. *Himara ne stuhitë e shekujve*. Tiranë: Shtëpia Botuese "ARBRI".
Verdery, K. 1994. The elasticity of land: problems of property restitution in Transylvania. *Slavic Reviews*, 53, 107–9.
Verdery, K. 1998. Property and power in Transylvania's decollectivization, in *Property Relations: Renewing the Anthropological Tradition*, edited by C. Hann. Cambridge: Cambridge University Press, 160–80.
Verdery, K. 2003. *The Vanishing Hectare: Property and Value in Postsocialist Transylvania*. Ithaca and London: Cornell University Press.
Vullnetari, J. 2007. Albanian migration and development: state of the art review. *IMISCOE Working Paper*, no. 18, September.
de Waal, C. 1996. Decollectivisation and total scarcity in high Albania, in *After Socialism: Land Reform and Rural Social Change in Eastern Europe*, edited by R. Abrahams. Oxford: Berghahn Books, 169–92.
de Waal, C. 2005. *Albania Today: A Portrait of Post-Communist Turbulence*. London and New York: I.B. Tauris and Company.
Winnifrith, T. 2002. *Badlands – Borderlands: A History of Northern Epirus/Southern Albania*. London: Duckworth.

Chapter 2

Against the Gated Community: Contesting the 'Ugly American Dream' through Rural New Zealand Dreams

Catherine Trundle

Between 2002 and 2004, in the picturesque Tasman region of New Zealand's South Island, a wealthy American migrant attempted to gain permission for a large beachside property development. Residents of the rural community soon organised a vocal campaign of opposition. Spokespersons claimed that the development would increase house prices beyond the reach of 'local people', and lead to an ostentatious 'gated community' in the midst of a population living on modest incomes. In this chapter I analyse the public outcry surrounding this controversy, placing it within the context of nationwide political debates regarding foreign land ownership and influence. Centrally, I explore how ideas of belonging, ownership and 'local' identity were deployed and made visible in ways that excluded a stereotyped version of the American migrant from the moral landscape. Moreover, I show how other American migrants attempted to align themselves with the opposition movement and distance themselves from 'rich Americans' and 'foreign' cultural values. Such stereotypes, I argue, acted as a foil through which 'average' American migrants could both personally and publicly confirm their rights to land custodianship, regional identity and virtuous migrant subjectivity. Such migrant claims relied upon discourses, also expressed by 'locals', of heritage, conservation, economic vulnerability and rural industry.

Based on fieldwork, conducted in 2004 for 10 months in the Tasman and Nelson regions,[1] this case study illustrates that contemporary mobile migrants may not embrace the transnational identity markers ascribed to them by social scientists. Their identities as migrants, furthermore, do not stem solely from their previous cultural worlds or their own personal aspirations and ideals. Migrants must also in various ways respond to the identity resources made available to them in their

1 During this time I kept field notes detailing informal conversations with residents and public discussion on the topic of 'rich Americans', collected all relevant local and national newspaper articles and letters to the editor on the issue, and conducted 15 in-depth interviews with American immigrants who had arrived in New Zealand in the last 10 years. All interviewees were of European descent. Nearly all had migrated while aged between 35 and 45 years old and just under half were in their 40s when interviewed.

new homes, visible in shared spaces and public discourses. Migrants might reject these resources or embrace them, but they must answer to them in some way. For the Americans of this study, identity work relied on localised motifs that explicitly countered and denied globalised and cosmopolitan frames.

An 'Ugly American' and a Beautiful Coast

In August 2002 regional and national newspapers reported that Californian immigrant Alan Trent, a permanent resident living in Tasman, had spent more than 12 million dollars acquiring several properties that made up a 200ha rural block along the picturesque Kina coastline (Collett 2002: D3). This sparked ongoing concern and protest amongst many residents describing themselves in public discourse as 'locals.' Localness is a complex and contextually deployed identity category (Savage, Bagnall and Longhurst 2005). In the context of this dispute, localness was claimed by New Zealanders of European descent living in these regions in order to assert a morally virtuous connection to place. Throughout this dispute, many vocal opponents of Trent made references to having been 'born and raised' in the Nelson and Tasman area, and many claimed their families had resided there for several generations. Such assertions of legitimate local belonging were often premised on a connection to colonial rural life, with many 'locals' emphasising their families' historical engagement with the establishment of the region's horticultural industry. 'Localness' in this post-colonial setting relies upon a settler ethos of being 'rooted in the soil' (see Lonsdale 2008). This connection was established by improving and civilising land through hard work, and of making it yield sustainable regional incomes and sociality.[2] That Trent was soon to be uprooting apple trees was thus powerfully symbolic of his emergent outsiderhood.

Locals' concerns were first publicly expressed in monetary terms, regarding the prices Trent was believed to have paid for the land. An article in the largest South Island newspaper, *The Press*, reported that:

2 Localness was thus complex. It did not just depend on being born and raised in the regions, or extensive genealogical ties to the area. Residents who had moved to the area and even some migrants felt they had managed to claim local identity and gain acceptance from 'born and raised locals'. These newer 'locals' were usually engaged in the horticultural industry, such as pip fruit or viticulture, and thus were engaged in local rural industry networks. Other groups, however, contested who should be included in localness. 'Alternative lifestylers' and 'hippies', for example, who had been living in the region for up to three decades argued that they too had their own connectedness to land based on 'self-sufficiency' and a spiritual engagement with nature. While the label 'local' is thus shifting and never morally neutral, I shall, for ease of expression, use it throughout the rest of the chapter to refer to those who self-define as 'locals'.

> Alan Trent first drew attention to himself when he bought his first Nelson property, at Kina, in 1999, reportedly paying $650,000 for a couple of hectares, well above what anyone else was offering for the land. In late 2000, he started his buy-up in earnest. Most of the properties he targeted were not on the market as such, but with pockets full of American dollars, and orchardists facing some of the most dire market conditions in generations of apple growing, he found willing sellers. He paid around valuation for some places, as much as double for several, and in pursuit of the final strategic piece in his new estate—41 hectares, including a swathe of cliff top costal land—he paid $3.34 million, nearly five times the rateable value. (Collett 2002: D3)

At this time, Nelson and Tasman were experiencing rapid increases in property prices.[3] Reports that 'rich Americans' such as Alan Trent were paying well above market rates for properties created an impression that foreign investors and immigrants, backed by the strong currencies of their home countries, were pricing New Zealanders out of the local property market. Active campaigners against Trent's proposed development believed that his actions were 'depriving ordinary Nelson people of the hope of ever being able to live there' (Collett 2002: D3). In a letter to the editor of the local paper a Nelson man wrote, 'He ... appears to be proud of contributing to a distortion of our economy, making it even more difficult for young families to buy their own homes. It is our country' (Smale 2003: 11). In such reports 'young' and 'ordinary' families became motifs of moral relationality set against the individualist economic ethos seen to drive Trent, and land became the symbolic and material battleground for debating what 'localness' meant and who it should include.

Concerns further increased when newspapers reported that Trent had a new plan to create a 200 house sub-division on his rural land, with the top properties priced up to 10 million dollars each (Collett 2002: D3, *The Nelson Mail* 2002: 1). Reports and rumours that this sub-division would be an exclusive and wealthy 'gated community' were met with considerable public unease (see for example Huband 2003b: 1, *The Nelson Mail* 2002: 1). Many locals feared that Trent's plans would result in a loss of prime productive rural land, and would transform a primarily rural and horticultural region into a residential (sub)urban region (Scollay 2002b: 1). This unease grew into vocal protests when Alan Trent began to build his own house and to landscape the surrounding property. The project

3 This can be seen in real estate figures between 1998 and 2004. The median property price in Nelson at the beginning of 1999 was NZ$143,500, and took an average of 50 days to sell. This did not fluctuate or grow by, on average, more than a 3 or 4 per cent until 2002. In the six-month period from November 2002 until April 2003 the median house price leaped by 25.3 per cent compared to the previous six month price median. The median price of a property was now $212,417, and they took, on average, only 17 days to sell (REINZ 2004a). The Tasman region experienced similarly rapid increases in property prices over the same period (REINZ 2004b, REINZ 2004c).

involved major earthworks and tree felling on his property's beachside cliffs (*The Nelson Mail* 2003b: 3). A residents' lobby group soon formed (Schollay 2002a: 3) protesting on historic and environmental grounds against the cliff earthworks and collected approximately 1,500 signatures petitioning the Regional Council to halt further alterations to the cliffs (Huband 2003a: 1).

The publication of an article in the *Los Angeles Times Magazine* in October 2003 exacerbated local fears further. It quoted Trent describing his proposed development to an American reporter.

> Trent has made New Zealand his permanent residence and has become a Kiwi citizen. Yet, his development is pure Californian. Dubbing his gated retreat Pebble Bay, a nod, he says, to California's Pebble Beach, Trent wants to re-create South Orange County, Dana Point in particular. And he wants to lure the 'Russell Crowes of the world' to buy into it ... The first home being built is his own. 'It's really nice, like something you would find in Malibu,' he says, 'an infinity pool, the whole Californian thing.' It is a high tech marvel with seven built-in plasma screen TVs and an elaborate multi-camera security system. 'It's kind of like a prison inside,' Trent boasts. In a small town with a crime rate that would be considered non-existent in America, why bother? 'It's just because you can do it, not because you need it,' he says, 'people with money, the Tom Cruises of the world, want security.' (Brown 2003: I12)

While the article was intended for an American audience it was circulated widely in the region via email. Summaries were published in nearly all major newspapers in New Zealand (Huband 2003c: 3, *The Press* 2003: 4, *The Dominion Post* 2003: 1, Sinoski 2003: 20, Dye and Trevett 2003: A4). Locals felt that their worst suspicions had been confirmed. Many people spoke of Trent's imperialist attitudes, as well as his desires to impose a 'fortified enclave' version of his 'American Dream' on his newly adopted community. Private concerns were now solidifying into a shared public discourse of oppositional identity work, illustrating how it is commonly through disputes that the markers of identities can become visible and consciously claimed (Banks 1996).

National Debates and the Public Control of Cultural Landscape

The unease that locals expressed regarding 'rich Americans' and their influence on the region fed into larger national debates concerning foreign control of public spaces. At the national level the sale of land to American foreign investors was making headlines and fuelling political debate. In order for non-residents to buy areas of land of any reasonable size and monetary value, investors needed to gain approval to make a Foreign Direct Investment (FDI) from the Overseas

Investment Commission (OIC).[4] According to the Commission's figures for land sales approvals in New Zealand from 1999–2003, investors from the United States of America ranked the highest, well over three times more than investors from the second and third highest land purchasing nations of Switzerland and the United Kingdom (OIC 2003: 12). In this same timeframe, the Nelson and Marlborough region were amongst the most popular areas for foreigners to buy land.[5]

The perceived permissiveness of the Commission towards these foreigners' applications, as well as the well-reported sale of 'iconic' areas of high country and coastal property to Americans, (see Johnson 2003, Laugesen 2003, Taylor 2003, Hodge and Patterson 2004) occurred at a time when national house prices were increasing rapidly (see REINZ 2004d). The combination of these factors led to a sense of public panic regarding the effect foreigners, and particularly Americans, were having on the cost of housing and land in areas considered to be 'commons' in New Zealand: the mountains and the beaches. Sensing that they were now competing on an international stage of unequal currencies, incomes and regulation, New Zealanders defining themselves as 'average' or 'ordinary' citizens began to publicly voice fears that such spaces were becoming privatised and commodified. This sense of geo-economic inequality was illustrated most starkly when those in positions of power within New Zealand were unable to compete for such spaces. Foreign investors were, in the public's mind, even excluding national elites and agencies of sovereign control from the marketplace. According to the nationally read *Sunday Star Times*:

> Overseas buyers often outbid the local millionaires—and even the government. The Natural Heritage Fund has about eight million a year to spend, says Forest and Bird president and fund member Gerry McSweeney, and it's just not enough. 'We are up against huge increases in property prices in the high country and on coastal properties, driven by overseas buyers.' The fund spent a quarter of its annual budget on Knuckle Point—and succeeded there only because the family trust that owned the land was prepared to drop the price below market value. A

4 According to the Overseas Investment Commission, 'the standard rule used to determine foreign direct investment in a New Zealand enterprise is that the foreign investor owns at least 25 per cent of the NZ enterprise.' (OIC 2001: 2) A foreign investor is classified as a non-New Zealand citizen, or a person who does not normally reside in New Zealand, a company formed overseas, and any company, trust or any other financial entity that is controlled or owned, 25 per cent or more, by overseas persons (OIC 2004: 2).

5 For example in 2003 and 2002 the region ranked third on the OIC table of 'land sales by geographic region' and in 2001 it ranked second (OIC 2003: 15). The total land sales approved by the OIC in the Nelson/Marlborough region from 1999–2003 was 22,090.4 hectares. In contrast the commission declined approval of 46 hectares of land in this region over the same time frame (OIC 2003: 11). Nationally, between 1999 and 2003, the commission approved the sale of 232,875 hectares of land to 'overseas persons', or 0.86 per cent of the total land area of New Zealand. In contrast it refused to approve consents for the sale of 1,838 hectares of land in total, or 1.6 per cent of applicants (OIC 2003: 7).

> determined foreign tycoon can easily beat the trust. Last year it lost a bidding war with American billionaire Julian Robertson for a prime slice of South Island high country. He paid $6.7 million for Brooksdale Station, says McSweeney—'three times the market price. He was in direct competition with us.' (Laugesen 2003: C1)

Such scenarios reflected the consequences of the government's increasing free market policies,[6] and the wider effects of a globalised neoliberal economy in which foreign investment must be encouraged with tax incentives and low regulation,[7] and where the increasingly fluid *mobility* of money clashes with more rigid notions of social *membership* (Jordan and Duvell 2003: 4–5). In such a context, immigrants and foreign investment taken together often become a potent symbol of danger, to the 'living standards, lifestyles and social cohesion' (Castles and Miller 1993: 13) of a geographically based group.

In 2003, the sale of land to 'rich Americans' became a tool for national politicking. Rod Donald of the Green Party, who called for more stringent restrictions on foreign land sales, argued that 'too much of our precious land has fallen into overseas ownership. Family farms are becoming a thing of the past as is the bach (holiday-house) by the beach because the price of land is being driven beyond the means of most New Zealanders' (Venter 2003: 1). Beachscapes, in such a discourse became a key metaphor for the creation of a nostalgic vision of the recent past, a fast disappearing era of egalitarian sociality in which nature was equated with public 'community' space.[8] As Bell points out, 'part of New Zealand's egalitarian tradition has been assumptions of shared access to publicly owned natural resources', such as mountains, parks, beaches and rivers (1996: 29, see also Curry 2001).

---

6 Beginning with the fourth Labour government of 1984, the New Zealand state has introduced policies of economic liberalisation which have seen New Zealand move from a heavily state regulated economy to one of the most internationally deregulated. See Castles, Gerritsen and Vowles (1996: 7–11), Easton and Gerritsen (1996: 22–47), and Kelsey (1995: 1–18).

7 The encouragement of FDI can also be seen in the relative ease in which FDI can enter and exit New Zealand, most notably through the absence of capital gains tax (Sinoski 2003: 20) on foreign direct investments. Another indication is the high number of consents approved by the OIC relative to applications rejected. The importance of foreign investment to the New Zealand economy is nowhere more pronounced then in the ranking of countries according to the 'significance of foreign direct investment in an economy'; in 2001 New Zealand ranked highest among the economies of developed countries for such investments (OIC 2001: 3).

8 See Arnold (1981: 260), Fairburn (1989: 12) and Belich (1996: 332) for a discussion of the colonial origins of the New Zealand egalitarian ethos and Pitt (1977: 5), Pearson and Thorns (1983: 234–44) and Wilkes (1994: 66–80) for a discussion of the cultural myth of the classless society within more recent New Zealand society.

Regular references to the 'bach' in this debate reinforced such an ethos. Iconic of the period between the 1950s to the 1970s, the bach was a beachside holiday house of only a few rooms, built and maintained by an 'average' New Zealand family through unassuming ingenuity and recycling rather than significant monetary investment (Peart 2009: 75–91). The bach in such moral imaginaries belonged typically to a family with children, and often multiple generations of family, rather than an individual. In terms of ownership, the bach thus signifies an inalienable space of custodianship that may be used (rather than definitively owned) by family members, but which ultimately expresses a communal ethos of shared engagement (2009: 85). Furthermore, as Peart notes, the typical 1970s bach was set back from the shoreline, allowing natural shifts in the shape of the coast to occur, and a strip of beach to remain for public access. By contrast, many new holiday homes are being built close to the water in order to ensure waterfront views. As the shoreline erodes, as has occurred in many parts of New Zealand, sea walls are being built to keep the ocean at bay. These walls have the effect of replacing costal strips of land and beach and thus denying public access to the area (2009: 126–7). The retro bach thus provides a model of shared familial engagement with nature and the commons that replicates up to the level of national relationality, identity and collective ownership.

Stereotyping 'Rich Americans' out of Local Landscapes

With neither the financial means nor the backing of local government, locals fought against the plans of Trent and other 'rich Americans' through the realms of public discourse and using the method of social alienation. These discourses utilised a series of moral binaries that cast Trent as the ultimate cultural and social Other, relying upon the tropes of rationalism versus affect, public versus private, and community/corporate versus individual. All such themes are visible in a resident's letter to *The Nelson Mail*:

> I feel strongly connected to a parcel of land in this region. My forefathers and mothers fled the famine in Ireland and came to this country to establish a new and better life. They purchased land and began farming and their descendants have continued to farm that land. I was privileged to live on that land, which has supported generations of my family. My sense of who I am is embedded in that land. The original farm remains intact. But for how long? District plans change, smaller and smaller parcels of land can be sold and wealthier and wealthier people are looking to stake a claim in this region. Many of these people don't have an understanding of the values and lifestyle, the shared memories and experiences, the shared losses and fortunes, the family sorrows and successes that have bound farming and orcharding families for generations. Apple trees may be easily uprooted, but the memories imprinted on land cannot be so readily dismissed. Money talks ... I object to the wholesale sell-off of the Nelson region

> to those who can afford it, to build huge haciendas for their 1.8 children. I don't want the development of Florida-style, walled, security mad, 'community,' with their own private golf, tennis and swimming facilities. I don't want the Nelson region to become some sort of playground for the if not idle rich, the anxious rich. I don't want Nelsonians to be in thrall to American, European or any other kind of major money … Fabulously wealthy immigrants come to New Zealand, buy up vast tracts of land which previously supported generations of families, drive up property prices and leave those who come from the area unable to buy a home … The same scenario is being repeated in many New Zealand communities. (O'Conner 2003: 9)

O'Conner's emotive letter, like many heated discussions of Trent among locals, conveys how the affective experience of 'locality' is not only emergent from the imaginative and cognitive realm (Appadurai 1996), but also embedded within place-based iterative action and relationality (see Amit and Rapport 2002). As Curtis has shown, notions of community often rely upon the fusion of 'emotional alliances and spatial associations' in ways that involve affective responses, embodied knowledge and emplaced sensation (2008: 404).

Yet evocations such as O'Conner's do not only convey established local identities, but actively work to build such identities. And such identities are not simply expressions of collective solidarity, but battles over resources (see O'Reilly, this volume). Like the symbolism utilised in proclamations of the endangered 'Kiwi bach', O'Conner paints a dystopic picture of wealthy immigrant space in contradistinction to Arcadian local space in order to claim ownership. In doing so she contrasts urbanity with rurality, and the destruction of nature with the creation of productive nature. Moreover, she asserts distinctions between contractive and divisive sociality (walls and shrinking families) and reproductive sociality, and the affect of upper class fear and anxiety in contradistinction to the affect of historically continuous 'community', memory and care. Journalists reporting on Trent's plans in the local newspaper reinforced and reproduced these oppositional identity claims. Trent was presented as having a rational ethos to land that lacked any spiritual, communal, familial or sentimental connections. In an article entitled 'Trent eyeing more land', Trent was portrayed as a socially disconnected individual viewing land solely as a commodity.

> [Trent] said he was not looking for any specific properties but would be generally in the market for a good investment. Mr Trent predicated that the property market would go even higher because land was underpriced to start with … He said he recently had a United States friend also wanting to buy land here because she is getting 1 per cent interest for her bank investment in California and, at age 73, wanted a better return for her money. She was looking at buying property here for her investment and hoped to achieve a return of 5 per cent. (*The Nelson Mail* 2003a: 3)

The newspapers furthermore subtly cast Trent as a social alien by focusing on the more extraordinary elements of his life. Trent complained in one interview that the media has sought to cast him as a 'nut' (Collett 2002: 3). The newspapers made repeated references to his membership of a 'small, independent evangelical church' and public rumours circulated, which Trent denied, that he was planning to create a 'Christian enclave' within his subdivision (Collett 2002: 3, *The Nelson Mail* 2002a: 1). As one Tasman resident quipped to me, 'we don't need a Nelson Waco here,' simultaneously invoking images of Millenarianism, isolationism and violence.

Trent's attitude to land ownership was further accused of undermining local democracy and heralding a new era of social inequality. Many residents felt that the democratic mechanisms of community consultation had been undermined by his property development proposal, and believed that the council was unwilling, or even powerless, to stop his development (Francis 2003: 13, McKenzie 2003: 7). Wealth and money was seen as a corrosive, morally dangerous force that had the potential to transform communitarian decision making processes into hierarchical tiers of political influence. District plans and consultation processes represented for many residents a form of common ownership, whereby absolute private property rights were mediated by the limitations imposed by a corporate 'community' body, represented in public discourse in socio-moral terms, as the 'public good'. Such a romanticised and idealised discourse had the effect of hiding forms of social differentiation and exclusion inherent within local relations and identity work for political effect. As one Nelsonian wrote in a letter to the editor of the local paper,

> New Zealanders do not need the … American Dream attitude, or the costal cliffs turned into fortunes. If it is not stopped now we will be like every country that has laid itself open to American influence—fenced off mansions for the rich, trailer parks for the workers and, like America, cardboard huts or doorways for the poor. Let the average American worker in, but let's keep 'New Zealand the way we want it'. (Madden 2003: 7)

The moral theme of egalitarianism can also be clearly seen in comments chosen by the Nelson media to be repeatedly reprinted from the above mentioned *Los Angeles Times* article. Another wealthy American migrant to the Nelson region, Glenn Schaeffer, was quoted saying that it takes a much smaller fortune to acquire fame in New Zealand than it does in the United States. The statement that drew the most public scorn was 'You hear my name up and down the country now … Cab drivers know me. You can be somebody [in New Zealand]' (see, for example, McGrath 2003: 13). When Schaeffer defended his comments, noting that unlike himself, very few people donate money to philanthropic causes in New Zealand, one Nelson woman responded, 'Obviously Mr. Schaeffer hasn't considered the options, that many prefer, to give anonymously or discreetly.' (Lenart 2003: 13) Entitlement to membership and ownership within the confines of this disputed rural space depended on forms of social conformity that emphasised solidarity, blending in, averageness and ordinariness. One's personal acts, effects on the built environment and engagement with political process were thus evaluated in public according to such principles in ways that excluded the stereotyped 'rich Americans' from local life.

Aligning with Locality

During this time, the Nelson and Tasman regions had become popular destinations within New Zealand for American immigrants.[9] Most of these migrants were not highly wealthy, yet all had to contend with stereotypes of 'rich Americans' in their daily lives. Americans that I interviewed believed that over the last several years anti-American sentiment had increased in the region. Most commonly, they attributed this directly to Trent's actions. Most participants mentioned reading letters to the editor or overhearing anti-American conversations and jokes. Participants described how the clearest marker of their origins – their accent – would often produce a negative response from locals. Brian recalled incidents when people arriving at his property to buy his products would 'hear our accents and they will be put off, because of everything that is going on around here … the Alan Trent thing.' Amy recalled how an American friend had 'Go home Yank' scratched on her car.

While acknowledging the difficulty of such experiences, most participants did not condemn such attitudes; many interviewees agreed with the fear expressed. In this sense, their interviews replicated local attitudes and dualisms, often in more emphatic terms than those used by locals. This allegiance can be explained with reference to the American participants' categorisations of American society and their previous lives. Nearly all American participants had previously resided in large cosmopolitan American cities. Despite having comfortable houses, extensive friendship networks, and creative, high-powered jobs, many explained that their lives were too stressful and busy, and they felt increasingly alienated from their own moral ideals regarding correct social interaction. They saw the United States as a dystopic space *par excellence*, in which work and money were valued above social relationships and family, and in which progress, industry and development had spoilt many rural and natural spaces, and the 'community' ethos that they contained. The decision to migrate thus involved the search for an Arcadian rurality associated in their minds nostalgically with a long past era.[10] 'Rich Americans' in

9 Out of the 74 districts of New Zealand, Nelson and Tasman respectively ranked 14th and 15th in terms of the number of North American immigrants gained over the period from 1993 to 2001 (Personal correspondence with Statistics New Zealand, March 2004).

10 While this search may involve a conscious rejection of American values, the search itself has roots in American culture. Nostalgia for a morally virtuous past is present in many areas of United States society. See McGirr (2001: 245–7) for a description of an American nostalgia for a bygone 'Golden Age' associated with the 1950s, close-knit communities and

New Zealand, many felt, represented a contamination, and all they had sought to leave behind.

Brian and Ella firmly distanced themselves from 'rich Americans' in two ways. They owned a vineyard and emphasised their necessity to work in order to live, stressing their lack of reliance on an overseas income. They discussed the value of working on the local land, the sense of 'connection' it gave them to the region and community, and the contribution their work made to 'improving' the regional economy and local social cohesiveness. Through such a narrative they expressed the morally correct ideal of productive and affective relationality with land. Furthermore, they emphasised their lack of geographic mobility. They described transitory wealthy American immigrants in New Zealand, those who fly between the USA and New Zealand in order to follow the summer, as 'snowbirds'. This, they explained, is a term used in the United States to refer to those Americans who travel south to states such as Florida in the winter months. When discussing these part-time immigrants Ella and Brian explained:

> E: When I meet Americans who say [She here put on a thick upper class American accent] 'Oh yeah we are only here for six months a year and then we go back to New England for the other six months', then I think, shame on you.
>
> B : What are you contributing [to New Zealand society]?
>
> E: Yeah, you know, coming here and living in your house in the summer and then leaving.
>
> B: That is what is different here, and that is ... [why] there may be a bit of animosity, because of that. All those people that just come and go ... are basically just taking the best from the region and not contributing [to it].
>
> E: Yeah, and they are here only at the best time [of the year].

While few of the American participants were snowbirds in this sense, most were mobile migrants, and returned to the United States once or twice a year, often for several months. They were also in high contact with friends and family in the United States via communication technologies. They could thus geographically and socially be defined as transnational migrants. Many American participants argued, however, that what distinguished their mobility from snowbirds, and negated its significance to their subjecthood, was the clear choice of one home that was foremost configured in terms of 'local community'. Discrediting multi-

'traditional family values'. While often a preserve of the political and religious Right, such an ideal has wider salience as an anti-modern trope. See also Hoey (2009) for a discussion of the importance of the 'Frontier' ideal in migration narratives associated with internal migrants 'downshifting' in the United States.

local loyalties was thus a fundamental narrative process in becoming the ideal type of American migrant. Geographically bounded, micro-level engagement was thus strategically asserted at the expense of national or international connectedness.[11]

Most American participants position themselves directly in opposition to Alan Trent and his controversial property developments in order to articulate localised belonging. Some had lodged submissions against his property development, and were engaged in groups seeking to improve community consultation. Referred to by one of the participants as 'King of the Hill', Trent was cast by most participants as the quintessential 'ugly American' immigrant. 'We're not trying to change this into Southern California, like Alan Trent is. And I understand the resentment, I don't want that either. I will fight him all the way', Peter explained. Similarly Amy told me at the start of the interview:

> You know that guy Alan Trent out on Kina peninsula? I want to knock on his door and just yell at him … I mean I understand: [in a sarcastic tone] 'Good on you, you have managed to make a wise business decision.' But it is at the expense of a community. That's what I don't agree with. I mean we could have all come out here seven years ago and bought the whole town, and especially Motueka. You could buy a section in town seven years ago for 20 grand, [or even] 10. And it is worth 80 now. And we all could have done that … But it would have been at the expense of the community … He came from America where every inch of coastal land is sold to someone wealthy and the houses are on top of one another … He comes from a culture where native bush is not valued like it is here … I don't know. I just feel very protective of New Zealand's natural environment. I think we can't have heaps of people coming here and doing all that. We just can't. And they are driving up the prices for the rest of us. I mean I will never ever be able to afford a bach now, ever.

Themes of environmental custodianship and conservation were crucial in many interviewees' assumption of a local identity and were sharply contrasted with Trent's attitudes. In such narratives, economic resources and environmental degradation were causally interlinked. Assertions of economic powerlessness were thus interwoven with claims to locality. Like Ella and Brian, Amy was now embedded within the regional economy. Earning the internationally weak New Zealand currency, she emphasised that her localness now prevented her from competing with 'rich Americans' for the ownership of iconic natural beach properties. A sense of victimhood and vulnerability to other migrants was crucial to the formation of a morally virtuous migrant identity.

---

11 In defining what made them feel at home in New Zealand, many thus talked about local landscape and place, rather than national icons or symbols, such a sports teams or famous figures. Furthermore, in discussion of politics, most issues centred on the Nelson and Tasman region, rather than the national political realm, or issues associated with other regions.

Such disputes were central topics of conversation in interviews, nearly always brought up by American participants without prompting. Laura, an American immigrant resident in Nelson for two years, expressed concerns over immigration policies and foreign investment issues in New Zealand that replicated local persons' core arguments:

> Because it doesn't serve you if you have got rich people … who are buying the best properties and are never here. And you have no neighbours. You have got these big empty shells. And when they come in, how do they support the community? How do rich people support us? … I would like immigration to take a whole 'nother look at that. Because you can come into New Zealand if you have four million [dollars], your application is [then] approved. How does a fact that someone can amass a large amount of money make them a prospective good resident of New Zealand? I think that criteria should be re-examined.

Migrants commonly typologise other migrants into ideal types in order to create a moral hierarchy of migrants (see Ip and Pang 2005, Benson, this volume). Such a discursive strategy maintains oppositional identity claims commonly directed at migrants, but internalises the debate within the migrant group, strategically shifting the boundary. Locals and 'good migrants' blur, while the stereotyped identity of 'bad migrants' solidifies into ultimate outsiderhood. Laura explained how her own publicly recognised contributions to local heritage had prevented her from assuming the 'rich American' identity.

> I have been so fortunate because of this house … Because when we bought it, it had been taken off the register [for] historic [buildings]. The people who owned it were working on having it carted away and having apartments built here. So when we bought the house and said no, it's not going to be carted away, this is going to be our home, everyone, I mean, I can't tell you how many people have come by the door and said 'thank you for saving the house.' So they didn't see us as Americans who were coming in to destroy the culture. They saw us as people who respected what was here … No, I respect anyone who is afraid of America coming here. I respect them, and their fears are well founded. I see America as the destroyer of cultures.

At the same time as showing how she evaded a 'bad migrant' identity, Laura simultaneously reinforces such a migrant stereotype as a fearful spectre threatening local life. As Cohen has observed, identity boundaries require a contrastive outside Other (1985, see also Barth 1969). What this case study illustrates, moreover, is that boundaries maintenance is a complex process: a group may be contested from both within and without; a migrant category may be split by migrants themselves; and the host society's identity may be less oppositionally deployed *vis-à-vis* migrant identity than public discourse initially suggests. Ultimately, the residence in and engagement with place constrains and provides opportunities for how migrant identity can be configured (see Curtis 2008).

Conclusion

Claims to locality are, in this case, strategies of migrant belonging expressed through specific arguments over place: how land and heritage should be preserved, how development should precede, how property laws should be legally and politically configured, and how people should relationally interact within a small geographic space. In asserting local identities, American migrants made use of a divisive migrant discourse that had contradictory effects. It gained them legitimacy in the eyes of many locals, creating a tiered typology of migrant types within some public debates. Yet it also added to fears of a broader migrant stereotype, contributing to the heated and confrontational debate about 'rich Americans', a source of anger and prejudice that locals at times directed back towards the American participants.

Many recent studies of migrants have focused upon how financially secure, mobile migrants creatively choose identities from a repertoire of available resources disembedded from localised place (for example Rapport 1997, Levitt, DeWind and Vertovec 2003, Fetchter 2007). Yet this study shows that American migrants in rural New Zealand made heavy use of locally available identity resources in the form of anti-immigration discourses circulating in a regional public sphere. These, after all, had the most moral purchase for making claims to belonging. Such claims were, however, not just instrumentally deployed to gain public acceptance, but integral aspects of migrant subjecthood that American migrants passionately expressed and acted out through daily practice. As part of their rejection of American life and their previous urban lifestyles, and their search for an Arcadian community space, local disputes were important resources in narratively justifying and reinforcing the migration decision and the binary categorisations of place upon which it rested. For many American migrants in New Zealand, who had been at the coalface of internationally mobile employment, urban living and the circulation of transnational money and cosmopolitan cultural capital, the reclamation of placed-based identity represented a critique of a particular American middle class way of life. Such a lifestyle creates as much disaffection and resistance as it does opportunities, even for those seemingly promised its best chances. In a world of movement, place has not lost its salience as a sphere for making meaning. Rather, it has re-emerged in unexpected realms of social life and is often, paradoxically, now intertwined with new modes of mobility.

References

Amit, V. and Rapport, N. 2002. *The Trouble with Community: Anthropological Reflections on Movement, Identity and Collectivity.* London: Pluto.

Appadurai, A. 1996. *Modernity at Large: Cultural Dimensions of Globalization.* Minneapolis and London: University of Minnesota Press.

Arnold, R. 1981. *The Farthest Promised Land: English Villagers, New Zealand Immigrants of the 1870s.* Wellington: Victoria University Press.

Banks, M. 1996. *Ethnicity: Anthropological Constructions.* London and New York: Routledge.

Barth, F. 1969. Introduction, in *Ethnic Groups and Boundaries: The Social Organisation of Cultural Difference*, edited by F. Barth. Boston: Little, Brown, 9–38.

Belich, J. 1996. *Making Peoples: A History of the New Zealanders from Polynesian Settlement to the End of the Nineteenth Century.* Auckland: Allan Lane.

Bell, C. 1996. *Inventing New Zealand: Everyday Myths of Pakeha Identity.* Auckland: Penguin Books.

Castles, F., Gerritsen, R. and Vowles, J. 1996. Introduction: setting the scene for economic and political change, in *The Great Experiment: Labour Parties and Public Policy Transformation in Australia and New Zealand*, edited by F. Castles, R. Gerritsen and J. Vowles. Auckland: Auckland University Press, 1–21.

Castles, S. and Miller, M. 1993. *The Age of Migration: International Population Movements in the Modern World.* Basingstoke: Macmillan Press Ltd.

Cohen, A. 1985. *The Symbolic Construction of Community.* London and New York: Routledge.

Curry, N. 2001. Rights of access to land for outdoor recreation in New Zealand: dilemmas concerning justice and equity. *Journal of Rural Studies*, 17, 409–19.

Curtis, J. 2008. 'Community' and the re-making of 1970s Belfast. *Ethnos*, 73(3), 399–426.

Easton, B. and Gerritsen, R. 1996. Economic reform: parallels and divergences, in *The Great Experiment: Labour Parties and Public Policy Transformation in Australia and New Zealand*, edited by F. Castles, R. Gerritsen and J. Vowles. Auckland: Auckland University Press, 22–47.

Fairburn, M. 1989. *The Ideal Society and its Enemies: The Foundations of Modern New Zealand Society, 1850–1900.* Auckland: Auckland University Press.

Fechter, A. 2007. Living in a bubble: expatriates' transnational spaces, in *Going First Class? New Approaches to Privileged Travel and Movement*, edited by V. Amit. New York and Oxford: Berghahn Books, 33–52.

Hoey, B. 2009. Pursuing the good life: American narratives of travel and a search for refuge, in *Lifestyle Migrants: Expectations, Aspirations and Experiences*, edited by M. Benson and K. O'Reilly. Farnham and Burlington: Ashgate, 31–50.

Ip, M. and Pang, D. 2005. Sojourners, model minority and multiple identities, in *New Zealand Identities: Departures and Destinations*, edited by J. Liu, T. McCreanor, T. McIntosh and T. Teaiwa. Wellington: Victoria University Press, 174–90.

Jordan, B. and Duvell, F. 2003. *Migration: The Boundaries of Equality and Justice*. Cambridge: Polity.

Kelsey, J. 1995. *The New Zealand Experiment: A World Model for Structural Adjustment?* Auckland: Auckland University Press.

Levitt, P., DeWind, J. and Vertovec, S. 2003. International perspectives on transnational migration: an introduction. *International Migration Review*, 37(3), 565–75.

Lonsdale, J. 2008. Soil, work, civilisation, and citizenship in Kenya. *Journal of Eastern African Studies*, 2(2), 305–14.

McGirr, L. 2001. *Suburban Warriors: The Origins of the New American Right*. Princeton and Oxford: Princeton University Press.

OIC. 2001. *Key Facts, Figures and Trends about Foreign Direct Investment*. [Online: Overseas Investment Commission]. Available at: www.oic.govt.nz/invest/brief/117511.pdf [accessed: 12 December 2004].

OIC. 2003. *Overseas Investment Commission Releases 2003 Figures*. [Online: Overseas Investment Commission]. Available at: www.oic.govt.nz/stats/annual_stats_2003.pdf [accessed: 12 December 2004].

OIC. 2004. *A Summary of the Overseas Investment Policies of New Zealand*. [Online: Overseas Investment Commission]. Available at: www.oic.govt.nz/invest/policies.htm [accessed: 12 December 2004].

Pearson, D. and Thorns, D. 1983. *Eclipse of Equality: Social Stratification in New Zealand*. Sydney et al.: Georg Allen Unwin.

Peart, R. 2009. *Castles in the Sand: What's Happening to the New Zealand Coast?* Nelson: Craig Potton Publishers.

Pitt, D. 1977. Are there social classes in New Zealand?, in *Social Class in New Zealand*, edited by D. Pitt. Auckland: Longman Paul Limited, 1–21.

Rapport, N. 1997. *Transcendent Individual: Towards a Literary and Liberal Anthropology*. London and New York: Routledge.

REINZ. 2004a. *Market Fact Graph, Results for Zone Nelson 1999–2004*. [Online: Real Estate Institute of New Zealand]. Available at: http://www.reinz.org.nz/reinz/public/market-information/market-information_home.cfm [accessed: 3 December 2004].

REINZ. 2004b. *Market Fact Graph, Results for Zone Motueka 1999–2004*. [Online: Real Estate Institute of New Zealand]. Available at: http://www.reinz.org.nz/reinz/public/market-information/market-information_home.cfm [accessed: 3 December 2004].

REINZ. 2004c. *Market Fact Graph, Results for Zone Takaka 1999–2004*. [Online: Real Estate Institute of New Zealand]. Available at: http://www.reinz.org.nz/reinz/public/market-information/market-information_home.cfm [accessed: 3 December 2004].

REINZ. 2004d. *Market Fact Graph, Results for All Areas [in New Zealand]* 1999–2004. [Online: Real Estate Institute of New Zealand]. Available at: http://www.reinz.org.nz/reinz/public/market-information/market-information_home.cfm [accessed: 21 February 2005].

Savage, M., Bagnall, G. and Longhurst, B. 2005. *Globalization and Belonging.* London et al.: Sage.

United Nations. 2004. *The World Investment Report 2004: The Shift Towards Services*. [Online: The United Nations]. Available at: www.unctad.org/en/docs/wir2004_fas_en.pdf [accessed: 12 December 2004].

Wilkes, C. 1994. Class, in *New Zealand Society: A Sociological Introduction* (2nd Edition), edited by P. Spoonley, D. Pearson and I. Shirley. Palmerston North: Dunmore Press, 66–80.

Newspaper Sources

Blowers, N. 2004. Letter to the editor. *Listener*, 27 March, 9.

Brown, C. 2003. It reminds them of California; Californians are leading a wave of immigrants to New Zealand, but some are encountering a Kiwi variation of NIMBY—not in my blue yonder. *Los Angeles Times Magazine*, 26 October, I12.

Collett, G. 2002. Buying into a storm. *The Press*, 24 August, D3.

Dye, S. and Trevell, C. 2003. Stroppy, poor locals irk US property investors: Californians living in New Zealand dish dirt on most beautiful country. *The New Zealand Herald*, 1 November, A4.

Francis, R. 2003. Letter to the editor. *The Nelson Mail*, 12 September, 13.

Hodge, K. and Patterson, C. 2004. A win for all New Zealanders: luxury resort halted. *The Dominion Post*, 2 December, 1.

Huband, B. 2003a. Trent defends activity on cliffs. *The Nelson Mail*, 22 October, 1.

Huband, B. 2003b. Trent 'bad guy' in US article. *The Nelson Mail*, 29 October, 1.

Huband, B. 2003c. Immigrant sought a Kiwi lifestyle. *The Nelson Mail*, 31 October, 3.

Johnson, A. 2003. Piece of paradise. *The Dominion Post*, 2 August, 1.

Laugesen, R. 2003. New Zealand for sale. *Sunday Star Times*, 2 March, C1.

Lenart, J. 2003. Letter to the editor. *The Nelson Mail*, 29 November, 13.

O'Conner, T. 2003. Not a wondrous feeling after all (Opinion piece). *The Nelson Mail*, 9 September, 9.

McGrath, P. 2003. Letter to the editor. *The Nelson Mail*, 22 November, 13.

McKenzie, A. 2003. Letter to the editor. *The Nelson Mail*, 4 November, 7.

Schollay, R. 2002a. Development sees group formed. *The Nelson Mail*, 16 August, 3.

Schollay, R. 2002b. Developer agrees to return with plan. *The Nelson Mail*, 27 August, 1.

Sinoski, K. 2003. No hard sell needed to tout NZ property to overseas investors. *The Independent*, 5 November, 20.

Smale, D. 2003. Letter to the editor. *The Nelson Mail*, 27 November, 11.
Taylor, K. 2003. Government plans crackdown on foreign land sales. *New Zealand Herald*, 11 November, A3.
The Dominion Post. 2003. California dreaming brings them to a cheap backwater. 31 October, 1.
The Nelson Mail. 2002. Costal homes to cost $10m. 14 August, 1.
The Nelson Mail. 2003a. Trent eyeing more land. 25 September, 3.
The Nelson Mail. 2003b. Ward calls for Trent's bulldozer to be confiscated. 8 December, 3.
The Press. 2003. NZ gets noticed in LA. 1 November, 4.
Venter, N. 2003. In foreign hands. *The Dominion Post*, 10 November. 1.

Chapter 3
Past Imperfect: Displacing Hawaiians as Hosts in a 'Drop Out' Community in Hawai'i

Lucy Pickering

Who is a host? What makes one group of people 'guests' or 'immigrants' and another 'hosts' or 'locals'? On the surface, this appears straightforward: migrants come from elsewhere and hosts are already there. But this simple guest/host distinction is not always so clear in practice. Drawing on ethnography with hippies and younger drop outs who have moved from the mainland United States of America to the Big Island of Hawai'i, I explore some of the implications of who guests choose to recognise as hosts. I focus in particular on one hippie, Dominic, who 'dropped out' of the University of Hawai'i and, as he saw it, mainstream American society in 1968 when he moved to the Puna district of the Big Island 'to grow 'erb' (cannabis), which he has been doing ever since. Through Dominic and his interactions with his Hawaiian neighbours, I highlight the importance of expectation in processes of host-making. Dominic and fellow drop out immigrants arrived in Hawai'i carrying particular (often New Age) assumptions and expectations about indigenous people, expectations left largely unmet in social life. Those Dominic expected to be idealised hosts turned out not to be, and so became for him just another set of colonisers in the islands. Dominic creatively reframed the concept of 'host' to take into account these unfulfilled expectations. It is these unmet expectations which lie at the heart of this chapter.

Interestingly, the category 'host' often remains largely unproblematised in studies of migration. The spotlight tends to focus on those who move. In the early years of migration studies assimilation with the 'host' society was assumed; however, as Levitt and Glick Schiller highlight, in recent years there has been a shift away from that assumption towards a greater recognition that some immigrants retain strong links with their home nation or live lives embedded in strong networks that span national boundaries (2004: 595). Out of this shift has emerged a focus on transnational identities, described through such organising principles as 'flexible citizenship' (Ong 1999), 'place polygamy' (Beck 2000), 'multilocality' (Rodman 1992, Trager 2001), 'cosmopolitanism' (Brooks 2000, Pollack et al. 2000) or 'hybridity' (Bhabha 1994).

Located alongside these is Hage's notion of 'existential mobility', where relocation in space is related to a sense of self as developing and moving. For

Hage, 'existential mobility' is not dependent on flow between places; rather, 'we engage in the kind of physical mobility that defines us as migrants because we feel another geographical space is a better launching pad for our existential selves' (2005: 470, see also Madison 2006 on his related concept of 'existential migration'). An analysis of migration which places 'existential mobility' at its heart does not depend on a construction of personhood which is hybrid, cosmopolitan or otherwise diffuse. Rather, it opens up a space for analysis which can operate with people who see the self, and see themselves, as more bounded and closed off; people who have consciously made a decision to leave their lives behind expressly in order to better launch their existential selves in a new space. For in 'dropping out', that is what these people have done: left one place and set of relations in order to find themselves in another. How did they find this remote corner of Hawai'i as a site of existential migration? In Sally's words, 'We just kept going west'; and in doing so they took with them the paradox of movement and rootedness in American national identity. This westward movement carries traces of past migrations west – to land both pristine and virgin and occupied by an indigenous Other: both *tabula rasa* and already populated place.

In his account of the United States of America's 'new elites', Brooks (2000) describes these elites as 'cosmopolitans': they are highly mobile, yet seek to consume – although not embody – a sense of rootedness to place through travel to rural United States backwaters and contact with families who have lived on the same land for generations. In his analysis, these cosmopolitans are playing out a central tenet of American identity – the tension between mobility and rootedness – and do so through their primary mode of engagement with the world, consumption. This tension is also played out in Counts and Counts' (1996) ethnography of North American RVers, who spend some or all of the year travelling around the continent in motor homes. According to Counts and Counts, this highly mobile segment of American society both typify and threaten an essential 'Americanness', for to be American is to be both rooted in place, living in community and knowing one's neighbours, and to be continually heading west, striking out alone and finding new territory. A common refrain among those drop outs who had travelled extensively across the United States mainland was that in coming to Hawai'i they had reached a natural end point of that journey. Hawai'i, rather than the road, was now 'home'. Existential mobility, whether in the form of continuous movement across the continent, dipping in and out of rural communities in order to feel rooted by proxy, or 'dropping out' and moving 'as far away from America as you can get without actually leaving' (Lisa), can be premised on different sorts of movements through space, but all remain tied to a notion of the self which develops as a result of that movement, and all play out a tension between movement and rootedness.

My research participants are migrants of a particular kind: they have 'dropped out' of their home society and their (predominantly) middle class upbringings, college educations and secure career trajectories in favour of the self-development that comes from relocating to rural Hawai'i and entering into the communes, cash-in-hand bodywork and cannabis trade networks which dominate locally. Some have travelled extensively before settling on Hawai'i as 'home', often having toured the country for years moving from festival to festival. Others, are like single mother of three, Sarah, who traded in life on the mainland for life in Hawai'i and explained that, 'I sold my trailer in Illinois, flew us all over here and bought our [1/3 acre] lot, because I might as well be poor where the weather's good.' Sarah's potential for further existential mobility now remains curtailed due to her renewed state of poverty, but at least it is poverty more on her terms. Some do retain economic links to other places, none of which are 'home'; for instance, Elsa dances tables in Alaska every summer and David trims 'buds' in California every autumn in order to finance their remaining nine months of the year in Hawai'i. All have moved to Hawai'i to 'drop out', although this could mean remaining in receipt of state-funded welfare, keeping in contact with family members on the mainland or dropping back in after a few months' sojourn in Hawai'i; or it could mean refusing benefits, losing contacts with mainland friends and family and living entirely within the locally well-developed illicit economy.

These drop outs are migrants: they have moved from one part of the United States of America to another, that is both the most westerly part of the United States of America and one far removed from 'home'. While some continue to move between Hawai'i and the mainland once relocated, such movement is uncommon and is rarely for purposes other than earning money. These migrants do not appear to experience Madison's (2006) existential crisis of being 'out of place' both in their original homes and in Hawai'i, and few articulate a desire, ambivalent or otherwise, to return. Nor are they cosmopolitan, articulating a sense of self based on movement and flow. Rather, they have made this out-of-the-way corner of a sparsely populated out-of-the-way American state 'home'. One of the ways in which those who can afford it (and many can) make this place 'home' is buying a piece of land.[1]

The ability to be able to buy a piece of land, own it outright and do with it almost what one wants is integral to an understanding of Hawai'i as an immigrant place. For a land of immigrants it is. Since Native Hawaiians have only recently become recognised by United States law as an indigenous population on a similar footing to Native Americans or Alaska Natives,[2] it is difficult to know the exact number of Native Hawaiians on the islands as they are still merged with other Pacific Islanders for census purposes. According to 2008 census data, 9.1 per cent of the people living in the state (11.2 per cent of those living on the Big Island)

---

1 My research participants fell into two broad categories in this regard: those who lived for a while – perhaps a few weeks, a few months or a few years – in one of the numerous 'communities', and those who bought land, usually between 1/5 acre and 5 acres in size, in one of the many subdivisions and attempted to build a new life here.

2 The Native Hawaiian Government Reorganization bill was passed by Congress on 23 February 2010 by a margin of 245–164 votes. Broadly speaking, it accords Native Hawaiians rights similar to those of Native Americans and Alaska Natives, although see Liermann (2001) and Kauanui (2005) for further discussion of the arguments put forward by sovereignty groups and others both in favour of and opposed to the legislation.

classify themselves as Native Hawaiian or other Pacific Islander; therefore over 90 per cent of the state's population categorise themselves as White, Black, Native American, Asian, Hispanic, or of two or more races; even taking into account this last group (18 per cent for the state, 25 per cent for the island), only a small minority identify themselves in a census context as either wholly or partially Hawaiian.

Long isolated from the rest of the world, an already populated Hawai'i was 'discovered' by Captain James Cook in 1778. Much of the indigenous population died out as a result of this contact with outsiders (although see Howe 2000: 64–8 for a critique of the 'fatal impact' hypothesis). Some of these outsiders hailed from the United States of America, arriving in the islands as missionaries and government advisors and underpinning the shift from the *ahupua'a* division of land (which was, in somewhat oversimplified terms, similar to European feudal land ownership in that land was commonly held rather than privately owned) to private ownership in the Great Mahele of 1848. Native Hawaiians often lost out in the Great Mahele, ending up with small or no land holdings, while immigrants from the United States bought up large tracts of land, and Hawai'i's sugar and pineapple plantation economy was born. To work these plantations successive waves of immigrants from around the Pacific Rim to as far away as Puerto Rico and Portugal were encouraged to relocate to Hawai'i, as free or indentured labour, in the late nineteenth and early twentieth centuries (see Kent 1983 and Merry 2000).

Out of this history emerged pidgin, the *lingua franca* of the islands today, and an ethnic identity, 'local'. To be 'local' in Hawai'i means to embody a history of (resistance to) various forms of colonial power and to speak the language born of it. With only one exception (a keen surfer who spent at least as much time hanging out with locals at surf spots as with drop outs in their homes and communes), none of my research participants spoke pidgin: all spoke Standard American English. Speaking different languages, eating different food, congregating in different places, drop outs kept largely to themselves and interacted little with locals in the area.

As a group, they position themselves in relation to locals and Hawaiians, and mainland retirees. These cultural categorisations were reflected in the three types of houses predominant in the area: hand-built 'hippie' shacks, American suburban-style bungalows and large and – by local standards – extravagant houses. According to my research participants, these were occupied by themselves, Hawaiians and locals, and *haole* retirees respectively. In reference to the latter – the large and ornate homes of white North Americans who had moved to the district for their retirement – my research participants were often critical of what they saw as the mis-match between house design and the conditions of the local environment.

For example, during the course of my fieldwork Brad, a well-known local cannabis campaigner and long-term hippie resident, lived in a compact, simple structure he had designed and built himself on a small but increasingly desirable subdivision along the coast. Two men commissioned a huge, brick (rather than wooden), unraised house to be built on the lot next door to Brad's. It filled the entire

1/5 acre lot and contained a washing machine, dishwasher and two bathrooms with flush toilets. Although the average annual rainfall for the region is 140 inches, there are periodic droughts when it may not rain for up to several weeks. There is no access to county water on that subdivision and, to Brad's amusement, their considerably larger catchment tank ran dry several times a year.

Dominic owns two 5-acre lots in a particularly out-of-the-way subdivision. On one of them he has a two storey tree house balanced on a series of cables strung between four trees (one of which is now dead), which he designed and built himself. As the salt water spray from the nearby ocean erodes these cables, Dominic is in a constant process of replacing them as they rust. A mainland retiree neighbour built a very large house right on the coastal edge of her lot, and used steel bolts to pin it together. As Dominic delighted in telling me, there's a strong probability of coastal shelf collapse along this stretch of coastline, so her house is doubly vulnerable: both to shelf collapse and to salt erosion of the steel bolts holding her house together.

While both Dominic and Brad would have much preferred different neighbours, both revelled in the ways in which these mainland retiree haoles did not understand the local environment, in this case the lack of a constant and predictable source of rain water and the corrosive nature of sea water. By comparing their own homes to those built by other mainland haoles, my research participants were able to show themselves to be different kinds of haoles, ones who were both independent (in designing and building their own homes) and in tune with the demands and restrictions of building and living in this environment.

However, while my research participants' homes were unlike those of 'other' mainland haoles, they were not necessarily like the houses of those whom they expected to best understand the local environment: Hawaiians and locals. There were large numbers of classic American suburban bungalows in Puna, often clustered together in particular subdivisions. Of the drop outs with whom I worked, I knew of only one uncle and nephew who lived in such a pre-fabricated home, which they were only renting while they built their own self-designed octagonal house on another subdivision with a higher proportion of drop outs. My research participants told me that these suburban pre-fab bungalows, usually on cleared and bulldozed 1-acre lots were occupied by Hawaiians and locals. Where possible, these houses were 'on grid' for water and electricity and, in my limited experience of the inside of these houses, contained electricity-draining and water-demanding 'mod cons' such as televisions, high wattage lights and washing machines.

The large houses of mainland retirees seemed to my research participants to be built with minimal understanding of the local environment. Locals and Hawaiians also seemed, with their perfectly flat lawns and identikit suburban bungalows, to be pursuing through their homes an American, not a Hawaiian, dream. To these drop outs neither other migrants nor the host population articulated an appropriate understanding of place, the demands of the local topography and climate or suitable use of resources. By contrast, Dominic and Brad both had small catchment tanks and were thus compelled to use their water in relation to rainfall – recent heavy rains meant long, luxurious showers; no rain meant brief or preferably no showers at all, and economical water use in washing up and cleaning. They both had solar panels attached to battery banks – a sunny day meant a little light or perhaps a film watched on Brad's computer after nightfall; a rainy day meant no lights and no films. They actively sought the restrictions of limited resources and saw in it virtue: the virtue of an enhanced sense of being 'in tune' with the local environment.

Such self-expression through building design is made possible only through the existence of private land ownership. Through the 1940s and 1950s, much of the largely unpopulated Puna district was bought up by developers, subdivided and sold sight-unseen on the mainland (Daws and Cooper 1990). While land ownership in Puna has been high since then, occupancy has remained low; thus it is easy for drop out cannabis growers to evade punishment by simply growing on unoccupied lots, and for Hawaiians to cross through such lots to reach the woods for hunting pigs or to access local fishing spots. It is this latter behaviour which led to conflict between Dominic and a group of Hawaiian fishermen who used his neighbour's vacant lot as a means of getting from the road to a popular moon fishing spot on the coast.

Shortly after starting fieldwork, I was sitting with Dominic one evening on the *lanai* (porch) of one of the buildings on his coastal lot, watching a group of Hawaiian fishermen troop noisily down through the bottom of his neighbour's lot to the fishing spot. I watched their progress with fascination, excited to see real fishermen out fishing, not when it was convenient, but when the fishing was good. I was intrigued by their actions, their conversation and that they would be going home that night with freshly caught fish. Dominic, who had sat on this *lanai* many, many more times than I, took a different view. He became immediately frustrated with the fishermen, not only because the noise and light they were generating disturbed our peace and solitude but because 'they always leave trash and I have to go and clear it up. And then I have to hitchhike into town with it to take it to the dump'.

A couple of weeks later I ran into Dominic in the village and, seeing that he was covered in cuts and bruises, I asked what had happened. 'I got into another fight with the Hawaiian fisherman,' he explained. Over a period of a several months, Dominic had become involved in a series of increasingly violent encounters with the fishermen, and with one in particular. In this most recent confrontation, the relatively small-framed Dominic described being picked up and thrown into the jungle. I myself never met this fisherman, however Dominic explained that the conflict revolved around access to the fishing spots and Dominic's persistent efforts to block them.

Dominic's neighbour's lot is unoccupied. Its owner lives on the West Coast of the United States of America and visits only occasionally. It is largely uncleared, although there is a crude path running most of the way down from the road to the coast, most of which an off-road vehicle can traverse; there are no buildings on the lot. There are 40 5-acre lots in this subdivision, with a thin strip of land running along the coast which is owned collectively by all subdivision residents. While occupancy is low (about 15–20 per cent), all lots are owned by someone. Thus this coastal strip, with its prime fishing spots, is only accessible to those who do not own lots by trespassing on someone's privately, but mostly likely absentee, owned land. Dominic's neighbour's lot was such a conduit.

The lights, noise and bags of trash annoyed Dominic and so, he explained to me, he asked the fishermen not to use his neighbour's lot to get to the coast. They continued to use the lot. He then explained that they were trespassing on privately owned land. They continued to use the lot. So Dominic rolled three large stones across the entrance, blocking vehicular access. The fishermen began walking from the road to the coast. Dominic contacted his neighbour to tell him that fishermen were using his lot as a thoroughfare and about blocking the entrance, which Dominic said his neighbour thanked him for. The fishermen continued to use the lot to access the coast, and as Dominic's efforts to prevent them were becoming ever more confrontational, so too were the encounters between Dominic and one fisherman in particular. Despite being beaten up numerous times, Dominic continued to protect his neighbour's private property, even though the fishermen merely passed through privately owned land to access a communally owned space which had long been used by Hawaiians, locals and haoles for fishing.

In discussing why he chose to be routinely beaten up in order to prevent these Hawaiians using his neighbour's lot to access this fishing spot, Dominic explained the situation on two potentially contradictory grounds. The first of these was that his neighbour *owned* the lot: they were trespassing on private property. That his neighbour did not use the lot was not as important as that the land that they passed through was privately owned by an individual who had not given explicit consent for these people, or indeed anyone, to cross his land. The second explanation was that Dominic's neighbour's land was one of six adjacent lots that Dominic was preserving intact for the Menehune.

There is some debate around exactly what or who the Menehune are; however for most of my research participants they were spirits who lived in the forest, particularly in the more sacred stretches of jungle, such as the one in which this subdivision is located. Sometimes they were experienced as fleeting glimpses between trees, but for most they were a felt presence when one was especially in tune with the jungle. For Dominic, they were spirits who spoke to him, telling him to preserve these six lots for them. This area was one of particular sacred importance to my research participants: a single road runs through it, unpaved and potholed and simply ending where the jungle has taken over to the point of rendering it impassable. The land on either side is filled with old-growth jungle where the guava has yet to take over, and there are numerous fruit trees lining the road itself, a consequence I was told of the decades, even centuries, worth of walkers throwing aside the seeds and stones of fruit they were eating as they walked. Three times there have been attempts to pave this road, each time popularly opposed, with the last attempt involving a hippie woman chaining herself to a mango tree. The Hawaiian bulldozer driver saw her attached to the tree, turned his bulldozer around and simply never returned. At the beginning and end of this stretch of road there is a sign in Hawaiian and on more than one occasion I listened to research participants recite the words as they passed the sign, unaware of what they translated as but reading them as a prayer, as recognition that they were entering a particularly sacred stretch of road. This stretch of jungle also lies exactly on 19.5° latitude, in line with the Green Mountain, also on the same axis: two sites connected by the power of sacred geometry.[3] The area also houses the highest concentration of 'communities' in the district. Attracted there by its energetic force, community members contribute to its spiritual power: balanced nature and human action acting in symbiotic feedback with the natural, sacred energy of the place.

Ivakhiv (2003) uses the term 'power places' to refer to those natural landscapes and ancient religious sites which are seen by New Age pilgrims as places that facilitate the channelling or concentration of energy. This stretch of forest is such a place, and a parallel can be drawn between people willing to tie themselves to trees to prevent the energetic balance of the jungle being upset by road paving and New Age opposition on the same grounds to the building of a bridge across Oak Creek, California (Ivakhiv 2003: 103). Ivakhiv highlights the tension over meanings and appropriate uses of places identified by New Agers as 'power places', between site use and site preservation. Instead of the debates around 'ritual litter' (Rountree 2006) and 'religious graffiti' (Ivakhiv 2003) which emerge out of New Age and Neo-Pagan uses of pilgrimage sites often shared with indigenous peoples, tourists, heritage managers, archaeologists and the like, this and other spots around Hawai'i were deemed by my research participants to have particularly 'high vibrations' but were not considered sites of pilgrimage as such. And whereas most debates around New Agers, Neo-Pagans and 'power places' concern their alteration of and impact upon places that others such as heritage managers want preserved intact, in Hawai'i these sites, and this stretch of forest in particular, were little used by anyone other than drop outs and fishermen.

Dominic's neighbour's lot was not the only one which people used as a shortcut to the ocean and on Saturday and Sunday mornings it was not uncommon to see bags of trash lining the side of the road; my research participants chose not to notice or discuss this trash, although when pressed speculated that it could only have been left by Hawaiians and locals since they themselves would always take their trash with them.[4] It was not the drop outs – who preferred a group ohm to a buried crystal – that changed these shared sites through their use of them, but rather indigenous others, who not only altered these sacred sites, but did so in distinctly profane ways.

---

3 The significance of 19.5° lies in it being the latitude at which the remaining points of a tetrahedron with one apex at the north or south poles make contact with a sphere's surface.

4 At the monthly full moon parties around the island it was notable that party sites were always left spotless the morning after.

One aspect of Ivakhiv's analysis of particular resonance here is his discussion of 'romantic Orientalism' (a term borrowed from Lopez 1995: 261, who in turn borrowed it from Said 1979: 154). For Ivakhiv, the New Age manifestation of 'romantic Orientalism' is the view that 'other' places such as India or Bali are more 'authentic' representatives of sacredness and spiritual wisdom. This view is not so dissimilar to Welch's 'Rousseau-tinted glasses' (2002: 21), which refers to the 'noble savage' vision of indigenous peoples, specifically Native Americans and Australian Aborigines, that she sees shared by New Agers and Neo-Pagans. Both authors draw attention to the tensions such an idealised vision of indigeneity can produce; as Ivakhiv argues, 'they are always expected to conform to the images "we" have created for them' (2003: 99). This is explored in greater depth by Welch who, like Ellen (1989), illustrates some of the many ways in which such romantic idealisations are misplaced, and goes on to examine the ways in which western media reinforce the noble savage narrative and how extensively this view is consumed by New Age practitioners. She offers illustrations of New Age practitioners understanding the current struggles of Australian Aborigines but choosing to ignore them in their promotional material and criticising Aboriginal speakers for talking of colonial oppressions: 'I came here to learn about your culture and to share in your culture [not] to be lectured at and made to feel bad about what happened in the past', complained one (Anon in Grossman and Cuthbert 1998: 779, cited in Welch 2002: 31). Welch's analysis does not limit itself to Westerners selectively consuming indigenous sacred acts/objects, but also looks at the many – not always complimentary – ways in which indigenous peoples interact with New Agers who are interested in their knowledge, and new media such as the Internet as a means of transmitting and controlling the transmission of culture.

The drop outs with whom I worked shared a similarly idealised notion of a globalised indigenous Other. They too valorised a vision of a homogenous, timeless Native American-looking Other. It was not uncommon to see t-shirts emblazoned with pictures of Native Americans beneath the banner 'Homeland Security 1492', posters around the village for shamanic workshops led by Americans who can trace their tutorage across Central and South America, or articles in the local free paper, read by nearly all my research participants, with titles such as 'Ancient Asian Wisdom for a Contemporary Society' or 'New Uses for Ancient Hawaiian Plants'. Ancient peoples and Native Americans in particular were the examples *par excellence* for gender equality, respect for nature, sustainability and peacefulness. Diversity and history appeared to have been effectively removed in favour of a straightforward, ahistorical, homogenous 'Rousseau-tinted glasses' image of an indigenous Other.

While less prominent globally, similar images of Native Hawaiians circulate in touristic and popular media. As Rennie (1998) reflects in terms of textual and Smith (1960) in terms of visual representation, the Pacific has long been imagined as an idyllic site. In relation to Hawai'i in particular, *Staging Tourism* (Desmond 1999) draws attention to the many ways in which Hawaiianness is softened, feminised, commodified and sold back to consumers as tropical idyll. Such images are not limited to touristic consumption, as Hawaiian sovereignty activist Haunani-Kay Trask demonstrates:

> As indigenous peoples, *our* nationalism is born not of predatory consumption, nor of murderous intolerance, but of a genealogical connection to our place … The wisdom of our creation is reciprocal obligation. If we husband our lands and waters, they will feed and care for us … This indigenous knowledge is not unique to Hawaiians, but is shared by most indigenous peoples throughout the world … We are stewards of the Earth. (1993: 128, emphasis in original)

This image of Hawaiians as 'stewards of the earth' does not sit comfortably with the local reality of Hawaiians driving souped-up trucks, spurning solar panels for 'on grid' electricity and leaving bags of trash in beauty spots. This is an incongruity which all my research participants were forced to confront in moving to an area with quite extensive Hawaiian Homelands resettlement, where the local indigenous people do drive, do drink and do watch TV; where they are not a timeless, homogenous, romantic Other. They dealt with this tension in different ways: some (a very small minority) became advocates for Hawaiian sovereignty, while many more chose to limit their social world to the clothing optional beach (rather than the surf spot), the health food store (rather than the Cash and Carry) and otherwise mix with other drop outs who shared their view of romantic Orientalism. In Dominic's case, he chose to regard the Menehune, mentioned fleetingly above, as Hawai'i's true aboriginal population.

According to the *Pocket Hawaiian Dictionary* the Menehune are a:

> Legendary race of small people who worked at night, building fish ponds, roads, temples; if the work was not finished in one night, it remained unfinished … Luomala 1951 convincingly refutes the theory that Hawai'i was settled by the Menehune before the arrival of the first Hawaiians.

Defined here primarily by their mythical monumental achievements, recognition is given – even if dismissed – to the argument that the ancestors of contemporary Native Hawaiians are not, in fact, the aboriginal population of the islands. While the Menehune do not appear in key historical accounts of this islands such as Fuch's *Hawaii Pono* (1961), Daws' *Shoal of Time* (1968), or Kirch and Sahlins' *Anahulu* (1992), they do appear in the short histories of tourist guidebooks. According to *The Big Island Revealed* (1999), when the Polynesians arrived they violently displaced and enslaved the Menehune population already living on the archipelago. The *Lonely Planet Guide to Kaua'i* says, 'According to legend, the menehune (little people) were Kaua'i's earliest settlers, who were displaced by subsequent Polynesian migrations' (Yamamoto 2006: 195). *Frommer's Kauai* (2004) explains that the non-Polynesian Menehune were displaced by Tahitian incomers who, being more aggressive and warlike, drove them north through the island chain to Kaua'i (2004: 193).

This sentiment is reflected in historian Joesting's 1998 *Kauai: The Separate Kingdom*, in which he argues that:

> In recent times the Menehune has been thought of as a group of mythical people, small in stature, who perform prodigious feats of construction during the night. It seems likely, however, that Menehune was the name given by the Tahitians to the early settlers of Hawaii who had come from the Marquesas ... The idea of a physically small race of people caught the imagination of Westerners and they made them into a colorful, romantic, and mysterious group of men and women who lived in the deep valleys and far up the mountain slopes. (1988: 20–1)

This, then, directly contradicts the *Pocket Hawaiian Dictionary*, which plays up their mythical status and plays down their historical 'reality'. In addition to these dominant but discrete discourses 'alternative' theories circulate which bring the two together, such as that of Childress (1998), in which the Menehune and Polynesians, be they Tahitian or Marquesan, co-existed and intermarried until the chief of the Menehune convened a night meeting to share concerns about being absorbed into the Polynesian people of Hawai'i. The next morning they were gone, either having disappeared on a floating island or retreated to 'remote inaccessible places' within Hawai'i (1998: 338).

To make matters even more complicated, it is worth noting the long history of the idea of a southern continent in the Pacific (see Rennie 1998 on medieval and later quests for a southern continent) and belief in an idyllic – and highly technologically advanced – Pacific island known as Lemuria, which, like Atlantis, disappeared long ago into the sea. As early as 1904, theosophists such as Scott-Elliott argued for the existence of a now-vanished Lemuria, populated with people technologically far in advance of us, or more recent contributors such as Joseph (2006), who argues not only that Lemurians were vastly technologically superior to us, but also that they left Lemuria to people other islands before its disappearance. My research participants believed in Lemuria, often arguing that it was a Pacific island populated by people possessing such advanced technological and spiritual qualities. Interestingly, a local variant on the Lemuria story was that cannabis first appeared in Lemuria, and radiated out from Lemuria to Hawai'i and elsewhere around the globe. According to some, it is even possible that Hawai'i *is* Lemuria.

In Lemuria, a parallel with the Menehune can be found: both were technologically and spiritually superior populations who were forcibly displaced and both are brought together with *homo floresiensis*, in Dominic's excited relaying of the news that a skeleton had been unearthed on the Indonesian island of Flores, shared one morning over breakfast:

> Lucy, they've found a skeleton in Indonesia or somewhere that's fully grown but only three feet tall. I think they might have found a Menehune and proof of Atlantis. You know, the peoples who built the phenomenal buildings in Cambodia and Mexico and Egypt, on Atlantis and Lemuria and here. Cos you've got to think about how these things were built, I mean you can't even get a credit card between the stones, and how did they get all the stones to fit so well together? It has to be because the Menehune could direct things with the power of their minds, you know, could affect the molecular structure of things by like getting them on an energetic level to turn into liquid and then gas. Then when they're in gas form they can take stones or whatever and move them to where they want them and direct them into moulds, which explains why you can't get a credit card between the stones, even though they are all different shapes … Of course, they're not around any more. To be honest, it was probably haoles chasing them out and killing them that wiped these far superior societies off the face of the earth. The Menehune still exist but not on our physical plane any more, they are on a spiritual plane at the moment.

The characteristics of Menehune, Lemurians and New Age 'romantic Orientalism' strongly overlap into one stereotyped indigene. Dominic's insistence that evidence of Menehune can be found across the Pacific and his easy collapsing together of Cambodia, Mexico, Egypt, Atlantis and Lemuria supports this conflation. Dominic, however, specifically referred to these aggressive incomers as haoles, not as Hawaiians. Translated from the Hawaiian, haole simply means 'foreigner', and in Pidgin stands for being 'generally arrogant, aggressive, ignorant of island culture(s) and histories, greedy, loud, and rude. Haole stubbornly continues to be 'foreign'; will not or cannot adapt to the island environment and culture' (Rohrer 2005: 92). Although commonly indexed by pale skin, the term haole carries a more complex set of meanings about outsiderness and rejection of what is already there. Relative to the small dark Menehune, Polynesians are paler-skinned foreigners who aggressively displaced Hawai'i's original, aboriginal, population. Seemingly unaware – or perhaps all too acutely aware – of the irony of he himself being one of a group of yet paler foreigners coming and culturally displacing Hawaiians, Dominic creatively reframes the ancestors of today's Native Hawaiians as haole: pale-skinned, foreign and 'arrogant, aggressive [and] ignorant of island culture(s)'.

Conclusion

The question of identifying who is a host, then, remains an important one. As people who have 'dropped out' of mainstream American society, and have participated in a form of 'existential migration', my research participants moved to a place with a particular host population in mind. They actively sought a 'Rousseau-tinted' pristine indigenous Other, only to be disappointed by the reality they experienced on the ground. This disappointment played out in their decisions about where to build a home (in subdivisions with a high concentration of 'drop outs'), and how to build it, both in terms of design and the act of building itself, as well as where to socialise and consequently who to socialise with. In Dominic's case, this was further played out through his protracted conflict with Hawaiian fishermen who did not use the land and ocean in ways that he deemed fitting.

Dominic's repositioning of Hawaiians as haoles enabled him to challenge the 'host' status of people who lived in suburban bungalows, drank beer, drove trucks and left trash in sacred 'power places'. For him, these Native Hawaiians, like yet paler-skinned American mainland haoles, were guests on the island, not hosts. As incomers, they too oppressed an indigenous people; they, like Dominic's colonial predecessors and retiree neighbours, behaved as bad guests. By identifying the Menehune as the true aboriginal people of Hawai'i, Dominic effectively sidestepped complex relations with and realisations about people who are generally recognised as Hawai'i's indigenous people. He also, equally creatively, sidestepped his own identity as a yet paler haole, and his own complicity in processes of (post)colonial domination. However, Dominic's repeated conflicts with the Hawaiian fisherman suggest that he could not completely sidestep either. His beliefs and actions may have helped him express a legitimate sense of belonging, but legitimate in the eyes of his fellow drop outs, rather than in the eyes of displaced Hawaiians.

Thus it remains essential to examine not only the multiple and complex ways in which migrants' identities and social lives are experienced as multi-local, hybrid or cosmopolitan, or the diverse ways in which migrants engage in place-making and localising, but also to look at who they do – and possibly more importantly *do not* – regard as hosts. In Dominic's rejection of Hawaiians as hosts and their replacement as such by the Menehune, the values, moves and role-models of 'New Agers' are revealed. But also revealed is a balance of power, both discursive and economic, in which some have the power to negate the host identity of others and those others are left debarred from engagement in long-practiced activities because of an apparent failure to conform to externally-imposed expectations of appropriate indigenous conduct.

References

Beck, U. 2000. The cosmopolitan perspective: sociology in the second age of modernity. *British Journal of Sociology*, 51(1), 79–107.

Bhabha, H. 1994. *The Location of Culture*. London and New York: Routledge.

Brooks, D. 2000. *Bobos in Paradise: The New Upper Class and How They Got There*. New York: Simon and Schuster.

Childress, D. 1998. *Lost Cities of Ancient Lemuria and the Pacific*. Stelle, IL: Adventurers Unlimited Press.

Counts, D.A. and Counts, D.R. 1996. *Over the Next Hill: An Ethnography of RVing Seniors in North America*. Peterborough, Ontario: Broadview Press.
Daws, G. 1969. *Shoal of Time: A History of the Hawaiian Islands*. New York: Macmillan.
Daws, G. and Cooper, G. 1990. *Land and Power in Hawaii: The Democratic Years*. Honolulu: University of Hawaii Press.
Desmond, J. 1999. *Staging Tourism: Bodies on Display from Waikiki to Sea World*. Chicago: University of Chicago Press.
Doughty, A. and Friedman, H. 1999. *Hawaii the Big Island Revealed: The Ultimate Guidebook*. 2nd Edition. Lihe'u, Hawai'i: Wizard Publications.
Ellen, R. 1986. What the black elk left unsaid: on the illusory images of green primitivism. *Anthropology Today*, 2(6), 8–12.
Foster, J. 2006. *Frommer's Kauai*. Hoboken, NH: Wiley.
Fuch, L. 1961. *Hawaii Pono: A Social History*. San Diego: Harcourt.
Grossman, M. and Cuthbert, D. 1998. Forgetting Redfern: aboriginality in the new age. *Meanjin*, 2, 771–880.
Hage, G. 2005. A not so multi-sited ethnography of a not so imagined community. *Anthropological Theory*, 5(4), 463–75.
Howe, K. 2000. *Nature, Culture, and History: The 'Knowing' of Oceania*. Honolulu: University of Hawaii Press.
Ivakhiv, A. 2003. Nature and self in new age pilgrimage. *Culture and Religion*, 4(1), 93–118.
Joesting, E. 1984. *Kauai: The Separate Kingdom*. Honolulu: University of Hawaii Press.
Joseph, F. 2006. *The Lost Civilization of Lemuria: The Rise and Fall of the World's Oldest Culture*. Rochester: Bear and Company.
Kauanui, J. 2005. Precarious positions: native Hawaiians and US federal recognition. *Contemporary Pacific*, 17(1), 1–27.
Kent, N. 1993. *Hawaii: Islands Under the Influence*. Honolulu: University of Hawaii Press.
Kirch, P. and Sahlins, M. 1992. *Anahulu: The Anthropology of History in the Kingdom of Hawaii*. Chicago: University of Chicago Press.
Levitt, P. and Glick Schiller, N. 2004. Conceptualising simultaneity: a transnational social field perspective on society. *International Migration Review*, 38(145), 595–629.
Liermann, A. 2001. Seeking sovereignty: the Akaka Bill and the case for the inclusion of Hawaiians in federal Native American policy. *Santa Clara Law Review*, 41, 509–46.
Lopez, D. 1995. Foreigners at the lama's feet, in *Curators of the Buddha: The Study of Buddhism Under Colonialism*, edited by D. Lopez. Chicago: University of Chicago Press, 251–95.
Madison, G. 2006. Existential migration: conceptualising out of the experiential depths of choosing to leave 'home'. *Existential Analysis*, 17(2), 238–60.

Merry, S. 2000. *Colonising Hawai'i: The Cultural Power of Law*. Princeton, NJ: Princeton University Press.

Ong, A. *Flexible Citizenship: The Cultural Logics of Transnationality*. Durham, NC: Duke University Press.

Pollack, S., Bhabha, H., Breckenridge, C. and Chakrabarty, D. 2000. Cosmopolitanisms. *Public Culture*, 12(3), 577–89.

Rennie, N. 1998. *Far-Fetched Facts: The Literature of Travel and the Idea of the South Seas*. Oxford and New York: Oxford University Press.

Rodman, M. 1992. Empowering place: multilocality and multivocality. *American Anthropologist*, 94(3), 640–56.

Rohrer, J. 2005. *Haole Matters: An Interrogation of Whiteness in Hawai'i*. PhD Thesis, University of Hawaii.

Rountree, K. 2006. Performing the divine: Neo-Pagan pilgrimages and embodiment at sacred sites. *Body and Society*, 12(4), 95–115.

Said, E. 1979. *Orientalism*. London: Routledge and Kegan Paul.

Scott-Elliott, W. 1904. *The Lost Lemuria*. London: Theosophical Publishing House Limited.

Smith, B. 1960. *The European Vision and the South Pacific 1768–1850: A Study in the History of Art and Ideas*. Oxford: Clarendon Press.

Trager, L. 2001. *Yoruba Hometowns: Community, Identity and Development in Nigeria*. Boulder: Lynne Rienner.

Trask, H. 1993. Malama 'Aina: take care of the land, in *Global Visions: Beyond the New World Order*, edited by J. Brecher, J. Brown Childs and J. Culter. Boston, MA: South End Press.

Welch, C. 2002. Appropriating the didjeridu and the sweat lodge: new age baddies and indigenous victims? *Journal of Contemporary Religion*, 17(1), 21–38.

Yamamoto, L. 2006. *Kaua'i*. London: Lonely Planet Publications.

PART II
Landscapes of Belonging

Chapter 4

'We are Not Expats; We are Not Migrants; We are Sauliacoise': Laying Claim to Belonging in Rural France

Michaela Benson

In this chapter I argue for a critical reconsideration of the significance of the 'local' as a category of belonging for transnational migrants. Drawing on material collected from an ethnographic study exploring the everyday lives of British lifestyle migrants living in the Lot, a rural, inland department in southwest France, I examine how the concept of locality is variously utilised and defined as they negotiate their lives in an unfamiliar setting.[1] While previous research on lifestyle migrants has stressed the extent to which these affluent incomers fail to integrate into local society (O'Reilly 2000, Oliver 2008), I present a more nuanced account of my respondents' efforts to become part of the local community, demonstrate the cultural context which gives rise to their ideas about locality and highlight the idea that their relationships with local social actors and compatriots are not mutually exclusive. For these lifestyle migrants, the desire for local belonging emerges as a driving force behind migration, a continued aspiration for post-migration lives and is intrinsically linked to the migrants' identity claims. As I argue here, the full significance of these claims to locality only becomes clear when they are examined within the context of the complex social relations that characterise their lives. Furthermore, the relationship of these claims to the wider rhetoric behind migration must be taken into account. Such claims to locality are, I argue, evidence of the complex social processes that my respondents have to negotiate in their lives following migration. What I examine in this chapter is not whether the migrants succeed in localising themselves, but what efforts to become localised signify both in the context of an increasingly globalised and interconnected world, and in relation to their ongoing quest for a better way of life.

My respondents' accounts and actions demonstrate both the desire and effort to become 'part of the community'. As the ethnographic vignettes presented below reveal, they are keen to stress their own efforts to become involved with the local

1 Department here refers to an administrative division in France, of which there are currently 100. The Lot covers an area of 5,217 km^2 and is one of the least populated departments (ranked 90th), with a population of 171,173 in 2007 (INSEE n.d.), and therefore a population density of 33 per km^2.

community, drawing attention to their relationships with local social actors and the land, and through their actions emplacing themselves in their new environment. These self-representations are a stark contrast to their negative representations of their compatriots also living in rural France. They mobilise these statements of belonging, not only as a form of self-identification, but also in their continued efforts, as members of the British middle classes, to distinguish their lives from those of their compatriots (Benson 2009). They continually engage in such processes of distinction, characteristic of the British middle class (Savage et al. 1992), as they strive to become indistinct from the local population, yet discrete from one another.

On one level the migrants' claims to local belonging are meaningful within the context of their aspirations for and lived experiences of life in the Lot, emerging as a discourse through which they achieve their goal of distinction. However, while it is clear that these claims are instrumental to the ways that they distinguish themselves from other Britons, they also need to be examined in terms of the cultural logic underlying them. Through the accounts of these migrants we learn that for them local integration has a moral value and lies at the core of the better way of life that they seek. They believe that by establishing and maintaining social relations with members of the local French community, they will be able to localise themselves. Social integration into the local community therefore emerges as a middle class aspiration and value (see Fog Olwig 2007, Torresan 2007). In contrast, integration into the expatriate community was presented as morally objectionable. It was thus common to hear the migrants denying their membership of such a community and downplaying their associations with their compatriots living locally, thus developing a strong rhetoric about the appropriate way to live following migration.

However, as the ethnography below demonstrates, rhetoric and practice are at times mismatched, with my own observations revealing that the migrants regularly misrepresented their relationships with other Britons. While it is not my intention to suggest that this in any way undermined their claims to local belonging, I question what these migrants' choices about how to represent themselves reveals. As I demonstrate, the migrants privilege their claims to local belonging and differentiate themselves from their compatriots to assert the distinctiveness of their lives.

The British in Rural France

In 2004 and 2005 I carried out 12 months of fieldwork among the British residents of the Lot. This was the first ethnographic investigation of the lives that these migrants lead following migration. Over the course of my fieldwork, I collected data through 75 in-depth interviews, informal conversations and participant observation. While my respondents shared common fatalistic presentations of life in Britain, they were by their own admission from all walks of life. They had migrated to the Lot in various waves (Benson 2010), with migration often coinciding with a watershed event such as redundancy or retirement (Hoey 2005, Benson 2010). Most of them had arrived in the Lot in the last 15 years, with several of them migrating only six months to a year before I conducted my research.[2] They ranged in age from 40 to 82, there were equal numbers of men and women and while many had migrated in couples or with family there were also some single individuals. They lived in disparate locations around the department, and while there were some areas in the west with large populations of British incomers, generally there would never be more than one or two British families in a village of around 200 residents. Finally, as I identify elsewhere (Benson 2010), despite their different life and migration histories, they were all members of the British middle classes.[3]

For my respondents, this middle class status could be read in the cultural determinants of their migration. Rural France was imagined as a rural idyll, an image that is particularly valued by the British middle classes (Williams 1973, Buller and Hoggart 1994, Barou and Prado 1995). Property ownership was also a central feature of their migration; most of my respondents had been able to sell their homes in Britain and become cash buyers in rural France. As Buller and Hoggart stated,

> French rural property ownership is becoming incorporated into the cultural experience of middle class British nationals. Even for those who do not own a French home, contemporary fiction, television and store catalogues increasingly allude to living in France. (1994: 127)

In other words, owning a French rural property can augment cultural capital. However, it seemed that this is inseparable from the experience of living in rural France; owning a French property was not enough to ensure this improvement, and migrants strived towards a sense of authentic living to qualify their social standing. In order to achieve this, they distinguished themselves from their compatriots, who they presented as leading less authentic lives (Benson 2009). As I have argued elsewhere (Benson 2009), they made these distinctions on aesthetic grounds, characteristic of British middle class processes of distinction (Bennett et al. 2009), which indicates the persistence of a middle class habitus.

More generally, the decision to migrate to rural France had been motivated by their belief that there had to be more to life than what Britain could offer. They did not move for work, nor were they forced into migration due to political or economic circumstances. Their accounts presented migration as a lifestyle choice,

2 There were individuals who had lived in the Lot for up to 50 years. As I relate elsewhere (Benson 2010), I had problems gaining access to these 'pioneers' who had disappeared into the local community, or who were unwilling to take part in my research.

3 I deliberately use the plural to denote the heterogeneity of the middle classes in Britain (see also Reay 2008).

undertaken to improve quality of life, and they can thus be counted as lifestyle migrants (Benson and O'Reilly 2009a, 2009b). They characterised Britain as lacking in community spirit, arguing that they never saw their neighbours, let alone had conversations with them, and that the ruthless pace of life was unforgiving to their health and well being. Their decision to move to the Lot was undoubtedly encouraged by their perception of rural France as a rural idyll with its slow pace of life, community spirit and beautiful scenery (Benson forthcoming; see also Buller and Hoggart 1994, Barou and Prado 1995). I argue that while these qualitative aspects of life in the French countryside were emblematic of the better way of life that these migrants sought through migration, the process of getting to this improved lifestyle continued long after migration and required a great deal of self-reflection (Benson forthcoming). In this respect, my respondents' quest for a better way of life could be understood as one articulation of a more general tendency toward self reflexivity within an era of late modernity (Giddens 1991), or as the pursuit of (elusive) happiness that Bauman (2008) argues is central to the experience of liquid modernity.

It is within this context of the migrants' ongoing quest for a better way of life that their claims to local belonging gain particular significance. For my respondents, expectations of life following migration seemed unanimously to pivot on their entry and acceptance into a vaguely defined, imagined local community consisting of French individuals and families who lived in the same villages as them. They valued the relationship of these natives to the land, and were keen to highlight the historical connections of these others to the Lot. In contrast, the local French population demonstrated a range of understandings about who could be considered a local or insider, and these reflected highly localised histories. For example, in one village that had experienced extensive depopulation to a critical point where the population would, without replenishment, die out, incomers such as my respondents were warmly welcomed. They were perceived as boosting the local population and were encouraged to take part in local activities. In other areas, ideas of what it meant to be local were contingent on having been born and bred in the Lot, as well as identification with the local imagined community. For example, in one village a farmer who had collaborated with the Vichy government during the Second World War while most of the other villagers had had sympathies for the resistance was never referred to in local terms and his identity was reduced to the German origins of his parents.[4]

My respondents would express their disappointment that they had not made as many French friends as they had hoped, stressing the difference between their expectations and their lived experiences. This mismatch between reality and aspiration was understandable. In their early expectations the migrants had a naïve

4 The Lot was in the Free Zone of France during the Second World War, and therefore governed by the Vichy regime rather than occupied by the Germans. There was an extremely strong resistance cell in the department, which is commemorated in the *Musée de la Résistance* in the department capital, Cahors.

vision of a traditional rural community characterised by mechanical solidarity, a cohesive whole (Rapport 1993) that would provide them with the antidote to the malaise resulting from their lives in contemporary Britain. They had hoped to find a place within this rural community and thus achieve the better way of life that they sought. However, they soon discovered that the rural idyll that they perceived in the Lot was not a cure-all; they found that French rural communities were by no means discrete, and even in the rural they could not escape their position as global social actors. Understood in this light, their persistent claims to local belonging spoke of their uncertainty about their position in the deterritorialised and globalised world.

Laying Claim to the Local in a Transnational, Globalised World

For my respondents in the Lot, claims to local belonging and locality reflected the key attributes of the better way of life that they sought. These are people who desired the comfort of the local, who wanted to feel rooted and whose migration could be described as a reaction against the perceived instability of the globalised world (see Harvey 1989). Accounts of other lifestyle migrants have focused on the construction of transnational communities running parallel to the host society (Rozenberg 1995, Rodríguez et al. 1998, O'Reilly 2000, Oliver 2008). These stress that prolonged engagement with the transnational presents an obstacle to the establishment of meaningful relationships with members of the local community. Authors such as O'Reilly (2000) argue that the rhetoric and actions of their respondents do not coincide; while they claim to be integrated, they are not, forming bounded communities that only include their compatriots. In this respect, it seems that many of these lifestyle migrants exclusively lead their lives within a transnational space, with their access to the local limited, not necessarily by the exclusions of the host population, but by their own actions.

In contrast, my own observations revealed that for several of my respondents in the Lot, claims to locality were more than just rhetoric; in fact, they visibly worked towards meaningful relationships with members of the local community. This ethnographic evidence suggests a need for an alternative conception of the lives of such transnational lifestyle migrants, which accounts for their relationships with their compatriots, but also for their efforts towards local community belonging. While I do not contest the continued importance of transnational connections in my respondents' daily lives, it is clear that the concept of the local was meaningful to their worldview and needs to be analysed accordingly.

Anthony Cohen (1982) wrote about the relationship between locality and belonging, arguing that subscribing to a particular locality is a form of (self) identification. Furthermore, his demonstration of the diversity of British rural cultures stresses that localities signify the boundaries between different groups of people. Just as Barth (1969) emphasised that ethnicity was an expression that responded to such boundaries, Cohen (1982) argues that locality, similarly an expression of culture, arises as an experience and expression of difference. It is through everyday, mundane practices that local distinctiveness is enacted, becoming more value-laden in the light of increasing threats to such ways of life. Correspondingly, he argues that ethnographic studies of locality, the focus of his edited collection *Belonging*, should provide:

> … an account of how people experience and express their difference from others, and of how their sense of difference becomes incorporated into and informs the nature of their social organisation and process. (1982: 2)

Cohen's focus on how individuals in small British rural communities experience belonging can perhaps help explain the value my respondents place on locality. With claims to locality acting to reinforce the boundaries between cultures, through their claims to local belonging my respondents articulated the difference between their lives before and after migration, thus distinguishing themselves from friends and families left behind. They sought to consolidate their sense of belonging through everyday practices and rhetoric. In this respect, it becomes less important to examine whether the migrants realise belonging and become part of the local community, and more pertinent to ask why the local is so meaningful to them and what their ideas and actions reveal about the cultural logic behind their quest for a better way of life. A general response to questions of the significance of the local to these migrants can be found within literature on globalisation. As many scholars have identified, there are an increasing number of links and interconnections between the local and the global (see Giddens 1990, Friedmann 1994, Kearney 1995, Inda and Rosaldo 2002, Lewellen 2002). The impact of this on anthropology has been felt in the rupture between culture and place; culture instead becomes deterritorialised and mobile, ready to be transposed on different cultural milieux (Gupta and Ferguson 1992, Hastrup and Fog Olwig 1997, Inda and Rosaldo 2002). In this rendering, locality can no longer be understood as spatially grounded (Gupta and Ferguson 1992, Appadurai 1996), and is instead constituted through relationships (Massey 1994, Hastrup and Fog Olwig 1997). Now understood as a property of social life, locality is continually (re)constructed (Appadurai 1996). It is therefore unsurprising that my respondents' claims to locality are primarily supported by accounts of their relationships with members of the local community. Claims to locality and claims to belonging thus emerge as interchangeable, both implying shared understanding and a sense of community, which may be imagined or actual. In this rendering, the importance of place to feelings of belonging is called into question.

Indeed, it is this understanding of belonging which supports Basch, Glick Schiller and Szanton Blanc's seminal work on transnationalism. Social fields stand where once locality might have been placed, in a representation of the dynamic ways that transmigrants negotiate belonging in a globalised world; 'many immigrants today build social fields that cross geographic, cultural, and political borders' (1994: 7). This understanding of the migrant experience encourages further examination of the relationships between migrants and those who do not move – both in the host and sending communities (Glick Schiller and Levitt 2004; see also Levitt 2003, Morawska 2003). In this rendering, claims to belonging are understood as statements about the individual's position within a circuit of relationships, and are not necessarily bounded by place.

This social fields perspective outlined by Glick Schiller and Levitt (2004) allows for the possibility of individuals engaging in multiple transnational processes simultaneously. The authors also rely on a distinction between 'ways of being' and 'ways of belonging', where the former refers to 'the actual social relations and practices that individuals engage in', and the latter to 'the identities associated with their actions' (2004: 1010). As a result, '[I]ndividuals within transnational social fields combine ways of being and ways of belonging differently in specific contexts' (Glick Schiller and Levitt 2004: 1010). My respondents could be understood as engaging within a transnational social field. The dynamics of the social field allow for their claims to belonging, where they self-identify as part of the local community, to be contradicted by their practices, which place them as part of a wider transnational community which incorporates their compatriots both back in Britain and living locally in the Lot.

Fundamentally, the question that remains unanswered in extant literature concerns the purpose of individuals' claims to local belonging. Appadurai (1990) has called for an investigation into the ways in which particular localities become meaningful to individuals and how these relate to their experiences within the wider globalised world. As Goldring argues, 'transnational communities and social fields offer a unique context for interpreting individual and collective status claims' (1998: 173).

The ethnography presented in the following two sections examines the migrants' claims to locality and their efforts to emplace themselves within the local community, while also recognising and discussing their underplayed relationships with one another. As I demonstrate, in order to understand why my respondents chose to privilege local belonging it is important to recognise that these identity claims took place within a wider social field of relations and were inherently caught up in their ongoing quest for a better way of life. Within this context, rhetoric and ideas about locality relied as much on the migrants' perceptions of life back in Britain as upon their experiences of rural French communities.

Self-identifications: Seeking and Realising Local Belonging

Claims to local belonging were significant for how my respondents identify themselves (see Cohen 1982), while also expressing their success en route to a better way of life (Benson forthcoming). The migrants' relationships with members of the local community and their desire to understand the landscape in local terms confirmed their efforts to integrate, but also reflected the migrants' own rhetoric about migration: leaving a Britain characterised by lack of community for the

Lot, a rural idyll abundant in community spirit.[5] The ethnography below traces their efforts to become part of the local community, demonstrating that unlike the lifestyle migrants in Spain who talk the talk but do not walk the walk (O'Reilly 2000, Oliver 2008), my respondents intended to realise local belonging.

The migrants' activities within local community life ranged from socialising with their neighbours to taking positions on the local events committee or teaching English in the local schools. These activities were explained as a way 'to be part of the village' or as 'giving something back to the community'. In the case of one particular village, accounts of the impact of these incomers stressed that they had helped revive the local community. The *commune* (administrative area) had a population of 92 people. They had lost half of their population during the industrialisation of France, which caused massive rural depopulation. In recent years, the local school had closed and as a result many young families had found themselves with no choice but to move out. There were fears that the population, which was largely elderly, would die out, and the history of the place would be lost.

Against this background, the recent swell in the population due to British immigration had been welcomed, particularly considering that the newcomers were eager to be involved in the community. At the time of my research, the British constituted 10 per cent of the population. They had, according to local social actors, rejuvenated the community, both in terms of age – several of them were in their 40s and 50s, while many members of the native population were aged 65 and above – and because their numbers helped to bring the *commune* back from the brink of extinction. These newcomers had an avid interest in local history, which they shared with their hosts, and seemed hungry for local knowledge. Members of the local French community recognised this and appreciated the migrants' efforts to learn about their new surroundings and get involved in local life. It was clear that the context of the village, which had experienced extensive emigration as a threat to community and was struggling to preserve its history and traditions, played a large role in the way that incomers were received. This transmission of knowledge was facilitated by the existence of a very vibrant community association, which regularly brought members of the local community together for meals, games evenings and general entertainment, as well as organising a rambling group.

In this respect, it was the migrants' contribution to a geographically bounded locality that helped shape their sense of local identity. As one of my respondents who lived in the area explained, 'We are not expats; we are not migrants; we are Sauliacoise', referring to his primary identification as a member of the *commune.* The migrants saw local identity, in this particular context, as being firmly bounded by place. Becoming part of the local 'imagined community' (Anderson 1983) was a realisable possibility. Local concepts of belonging were in the process of changing

5 Lack of community in the home country is a frequent complaint among lifestyle migrants, and a common reason given for why they chose to migrate in the first place (Benson and O'Reilly 2009a, 2009b).

to incorporate the contributions of and commitments made by these incomers. Therefore, in laying claim to this particular locality, the migrants engaged with their local friends in a shared sense of the community and its future.

Establishing relationships with members of the local community also motivated some of the migrants to join sports teams. Unlike the cricket teams set up in the Dordogne (a neighbouring department) by British incomers, these are clubs that have been set up by the local French. Alice Hammond recalled how she had encouraged her husband Keith to join the football team in their village because she saw this as a good opportunity to make friends.[6] Although Keith initially spoke no French, his regular attendance at practice and general support for the team eventually broke down the resistance he had felt from some members of the team, and he established lasting friendships. Alice was also able to make friends with the other wives and girlfriends of the players by attending the team's matches and accompanying Keith to social events. The local French people clearly recognised and appreciated the effort that Keith and Alice had made, commenting, '[Y]ou and Keith just aren't English. Are you sure you're not French? You and Keith are different'. Here, 'English' was presented as a point of reference, a stereotype of the people who come to France and don't even try to mix with the local population. It was clear that the commitment that Alice and Keith demonstrated through this regular participation facilitated their inclusion into localised networks.

Of course, involvement in these clubs was contingent upon a certain level of fitness and competence, which excluded some people from taking part. Others with a love of sport but beyond an age where they could play would go and watch live games, and had made friends with like-minded people this way. In all of these cases, initial contact was made on the grounds of a shared interest, but the relationships then had a chance to develop further. For example, William and Victoria explained that they had first met their French business partner through their participation in the local rugby club.

However, there were cases where this strategy was not as successful. It seemed that there were fewer opportunities for my female respondents to take part in sporting clubs. Every *commune* had a keep-fit club, as the local actors referred to it, but in my own experience this was often an informally run affair attended mostly by old(er) women, and its purpose seemed to revolve around gossip rather than any real desire to get fit. While this could be considered the perfect opportunity for socialisation, my respondents who had attended explained that in the early days they had found the language difference and lack of common interests a barrier to inclusion. Trying to have a meaningful discussion while jogging breathlessly around a small, cold room was not conducive to making friends. Connie explained to me that she had attended the local keep-fit club once or twice, but had been put off by the mindless gossip and the lack of rigorous exercise. In this respect, Connie's experience at the keep-fit club was a reminder of the differences between

6 All names appearing in the text are pseudonyms.

herself as an incomer and the locals. Perhaps she felt excluded, or that their way of relating to one another did not live up to her expectations.

It seemed that for my respondents gaining knowledge of their new surroundings from the local population further mediated their relationships, in turn perpetuating the sense of community they sought. The village rambling clubs provided such opportunities.[7] Guided by their French neighbours, the migrants learned local history and heard stories about the area; this helped them begin to see the landscape through the eyes of the locals. For example, Martin and Sarah Johnstone recounted their walks with the local rambling club, which had often been guided by an elderly French man. As they walked, they learned not only the history of the landscape but also local mythology. Their excitement at learning that one of the fields contained small earth mounds that had been created by a local magician searching for gold demonstrated their desire for this particular type of local knowledge, through which they came to understand the relationship of the local people to the land. The local knowledge that my respondents gained was significant to their claims to local belonging; it was important that they had gained this knowledge not through reading books, but through their relationships with members of the local population. In this respect, the migrants' knowledge of their surroundings was less important than the process by which they gained this knowledge.

Walking through the landscape was important in shaping the migrants' ideas about local belonging, as were their other phenomenological experiences of the land. Following migration, they experienced an increased engagement with the landscape as they set to work planting their vegetable gardens. This was no mean feat; they battled against the rocky earth as they dug and ploughed the ideal vegetable patch and they fought the elements as they tried to save their crops. In this manner, the migrants attempted to emplace themselves in the Lot, learning about their new environment through their embodied practices. Ingold (1993, 1995) has argued that it is through these engagements with the landscape, alongside their relationships with local social actors, that individuals localise themselves. But there is a further element of these engagements that is pertinent to the migrants' performances of local belonging: by presenting their struggles with the land, they implied that they had grown to understand it as their French neighbours did, as something to be tamed (see Theodossopoulos 2003).

In contrast to theories that emphasise the importance of social relationships in constructing a sense of locality, for many of my respondents in the Lot both social relationships and place together contribute to their sense of local belonging. Proximity and embeddedness in the local environment are thus mutually constituted. The sense of community that my respondents imagine and work towards is physically designated, with the relationships between members reliant on geographic proximity. Despite their individual biographies as transmigrants they favour a relationship between place and identity, similar to that found in many

7 Almost all villages and towns in the Lot had a rambling club, although these met with varying success.

of the cases presented in Cohen's (1982) collection. Whether they had achieved it or not, the migrants recognised that for community to be realised, they needed to work towards it, make a contribution and show the local people that they were interested in their lives.

'Being' Exposed: Distinction as a Transnational Process

Despite the migrants' efforts towards the realisation of local belonging, my detailed examination of their lives demonstrates that they simultaneously maintained connections with a selection of their compatriots living locally and back in Britain. On the one hand, their tendency to underplay the importance of transnational relationships in their lives revealed a moral stance favouring integration and their own efforts to become part of the local community. On the other hand, this tendency made a statement about the improved quality of their lives, thus justifying their migration. In this manner, they distinguished themselves both from invading incomers – such as colonialists or lifestyle migrants in Spain and the Dordogne (see Benson 2009) – who imposed their own cultural norms and values onto their new surroundings, and from their friends and families left behind, who appeared powerless to change their lives despite their discontent with life in Britain (see also Trundle, this volume). In stark contrast to their positive self-presentations as actively seeking relationships with members of the local population and avoiding the expatriate community, the migrants put forward negative representations of their compatriots living in the Lot. As I discuss elsewhere (Benson 2009), it was common for my respondents to stereotype their compatriots as a generic group who did not attempt to establish relationships with members of the local population, could not speak French, drank too much, spent too much time with one another and maintained too many links with their lives back in Britain. At other times, they did not hesitate to name compatriots who they felt were not approaching their lives in rural France appropriately.

Through discussions of these 'others', the migrants enhanced their presentations of themselves as distinctive. They strengthened these claims by demonstrating the extent to which their lives were different, emphasising the meaningful relationships that they had with local social actors and their (local) knowledge of their new environment. Their continual references to the way in which they were different to others, and their goal of a different way of life resonate with Bourdieu's (1984) argument that distinction is characteristic of middle class engagement in class reproduction (Benson 2009). Through this process, they accumulate cultural capital (Bourdieu 1984, 1986, 1996).

It might be assumed that, given their efforts to demonstrate that their lives in rural France were different to those of their compatriots, my respondents would avoid contact with one another. Indeed, this was the rhetoric employed by some to explain why they had not joined the local branch of Association France Grande-Bretagne (AFGB), a bilingual cross-cultural association, stressing that they felt it was now more British than French.[8] However, to argue that the migrants did not have any contact with their compatriots would be a gross misrepresentation; their friendships with one another remained a consistent and important feature of their daily lives. While they would refer to a 'British community' that they were somehow external to, it became clear that at times they took some comfort in the company of these other expatriates.

My own investigations suggested that all of my respondents were connected to each other through at least one mutual friend, and that this was probably representative of the relationships between Britons living around the department. They met each other at French lessons and at the AFGB, which several of my retired respondents had joined, but they also met randomly in the waiting rooms of doctors' surgeries and the hospital, waiting in line at the post office and on market day at the Interlude, a popular café in Cahors. While their relationships with local social actors often took place within a localised space, their interactions with one another took place in a wider geographical space, reflecting their own travels around the department to visit the best English-speaking doctor, to attend French language workshops or to visit the nearest post office. Within this circuit a certain amount of knowledge about how to live in the Lot – where to buy certain products, which restaurants to visit, which doctor to go to, new laws, myths about French bureaucracy and so on – circulated among the British residents of the Lot. In this respect, among their compatriots living in the Lot such knowledge became a marker of integration and embeddedness. It demonstrated their capability to negotiate their new environment successfully, and was a source of symbolic capital.

It is evident that, despite their protestations, the migrants found their relationships with their compatriots comforting and a source of emotional support. One of the younger migrants, who had had a nervous breakdown shortly after migration, attributed her recovery in part to the support of her British friends in the area. 'We're like family', explained Connie, who had lived in the Lot for 12 years, stressing that their compatriots living locally could be called on in times of need. It was Julian who Harry called when he needed help building a pond in the garden, and Vic who he called when he found himself housebound and his wife Connie needed a lift. Pat and Jean Porter, who ran a vineyard, annually called upon a group of British volunteers to help them handpick the grapes for their wine, explaining that these were their only acquaintances available at that time of year; all of their French neighbours were busy on their own farms and vineyards. Nevertheless, accounts of their relationships with other British migrants were often fleeting, as though they did not want to draw too much attention to them, fearing that they would undermine their rhetoric of integration.

Equally, it was also common for the migrants to insist that they rarely travelled back to Britain. This signified a further rejection of their transnational belonging and was supported in practice. For example, several of the retired men refused

8 At the other end of the spectrum, I met some people who had chosen to join the association precisely to meet other Britons living in the area.

to leave France even to visit their children, allowing their wives to travel back alone. Other migrants could count the years it had been since they had last visited, stressing that 'there is nothing for us to go back for'. Those who still had family obligations were adamant that without these they would never return. It seemed that my respondents were uncomfortable with their simultaneous identification with Britain and their burgeoning commitment to local life in France. Yet they continued their weekly telephone calls to friends and family back in Britain, and with the arrival of email there were new possibilities to maintain their transnational connections. The Freeview® television boxes that many of them had also allowed them to keep up to date with national and local news from Britain, opening a communicative space for themselves, their friends and families back in Britain and their compatriots living locally. Many also regularly had friends and family visiting them in their homes in the Lot.

Against this background, the migrants' discussions about 'others' should not be taken at face value. My respondents did not so much reveal the intricacies of other people's lives as reflect their own aspirations and the way that they wanted to be seen. Despite their apparent ambivalence, their relationships on this transnational level remained a feature of their daily lives. These actions and practices, 'ways of being' as Levitt and Glick Schiller (2004) refer to them, are clearly at odds with the rhetoric of local belonging that the migrants readily project. However, as the recent work on transnationalism has demonstrated, 'ways of being' and 'ways of belonging' do not necessarily have to correspond with one another (Levitt and Glick Schiller 2004).

Conclusion

The mismatch between being and belonging begs the question: what do declarations of local belonging and misrepresentations of their other social relations do for my respondents in the Lot? On one level, it is evident that their claims to the local act as purposive self-identifications that are made meaningful in contrast to their lives led before migration, the lack of community spirit in Britain contrasted with the vibrancy of local community in the Lot. Within the context of the migrants' ongoing quest for a better way of life (Benson forthcoming), claims to locality served as indicators of their progress, reinforcing and justifying the decision to migrate and their continued residence in rural France. However, on another level, the value placed on local belonging was also implicitly moralising. While they presented themselves as benign and considerate lifestyle migrants who embrace the local community, those who failed to work towards integration were denigrated. By recognising an underlying cultural logic in which integration and transnational belonging were structurally opposed, it became evident that claims to locality operated within the processes of distinction in which my respondents regularly engage (Benson 2009).

The quotation 'We are not expats; we are not migrants; we are Sauliaçoise' highlighted that it is through their insider status that my respondents were able to claim a distinctive way of life. Following Herzfeld (1992), I argue that they were different precisely because they felt that they were integral members of the Lotoise population. And through their claims to local belonging they distinguished themselves from other migrants who had not been so successful. Inspired by Bourdieu (1984), I have argued elsewhere that the processes of distinction that characterised my respondents' lives draws attention to the extent to which their middle class status continued to influence their lifestyle choices (Benson 2009). Indeed, as Fog Olwig (2007) and Torresan (2007) demonstrate, settlement and integration can be understood as the continuation of middle class aspirations. For my respondents in the Lot, the achievement of a better way of life rested upon such integration. While they stressed their anxiety over how the local French see them, keen to pre-empt any suggestion that they were taking something away from French rural life, I argue that their rhetoric of local belonging was also meaningful in what it conveyed to their compatriots. Indeed, the recognition of improvement in their lives came not from their French friends but rather from other Britons.

Through migration and property ownership the migrants had the possibility of enhancing their cultural capital (see Bourdieu 1984). However, this was not a straightforward exchange, with success determined by other lifestyle achievements, namely integration. This reveals the moral value that British middle class migrants placed on integration, a response perhaps to their perception that their working class compatriots did not integrate. In this rendering, they avoided presenting anything that could call into question their claims to local belonging and risk the related accumulation of cultural capital. This explains their efforts to conceal the extent of their relationships with their compatriots.

This chapter, alongside the other contributions to this volume, has demonstrated that it is no longer possible to ignore the local, even when exploring the lives of transnational migrants. For my respondents in the Lot, ideas about the local inspired the decision to migrate and influenced their expectations and experiences of life following migration. Nevertheless, the migrants' claims to the local were made meaningful in the context of their engagements in a circuit of social relationships extending beyond the local (Glick Schiller and Levitt 2004). While they underplayed their relationships with their compatriots, it was clear from their actions that they relied upon these as they came to terms with their new lives in the Lot. Furthermore, these relationships were intrinsic to the migrants' project: the search for a better way of life. Through their interactions with their compatriots they confirmed and reinforced their decision to migrate, assessed their position en route to a better way of life and sought authentication of their post-migration lifestyles. As the chapters in this collection demonstrate, it is no longer possible to take claims to locality at face value; they exist within a circuit of relationships in which contemporary migrants regularly engage, and which have an impact on how they are experienced, utilised and understood.

References

Anderson, B. 1983. *Imagined Communities: Reflections on the Origin and Spread of Nationalism.* London: Verso.

Appadurai, A. 1991. Global ethnoscapes: notes and queries for a transnational anthropology, in *Recapturing Anthropology: Working in the Present*, edited by R. Fox. Santa Fe, New Mexico: School of American Research Press, 191–210.

Appadurai, A. 1996. *Modernity at Large: Cultural Dimensions of Globalization.* Minneapolis: University of Minnesota Press.

Barou, J. and Prado, P. 1995. *Les Anglais Dans Nos Campagnes.* Paris: L'Harmattan.

Barth, F. 1969. Introduction, in *Ethnic Groups and Boundaries: The Social Organization of Culture Difference*, edited by F. Barth. Illinois: Waveland Press, 9–39.

Basch, L., Glick Schiller, N. and Szanton Blanc, C. 1994. *Nations Unbound: Transnational Projects, Postcolonial Predicaments and Deterritorialized Nation-States.* London: Routledge.

Bauman, Z. 2008. *The Art of Life.* Cambridge: Polity.

Bennett, T., Savage, M., Silva, E., Warde, A., Gayo-Cal, M. and Wright, D. 2009. *Culture, Class, Distinction.* London: Routledge.

Benson, M. 2009. A desire for difference: British lifestyle migration to southwest France, in *Lifestyle Migration: Expectations, Aspirations and Experience*, edited by M. Benson and K. O'Reilly. Farnham: Ashgate, 121–36.

Benson, M. 2010. The context and trajectory of lifestyle migration: the case of the British residents of southwest France. *European Societies*, 12(1), 45–64.

Benson, M. Forthcoming. *The British in Rural France: Lifestyle Migration and the Ongoing Quest for a Better Way of Life.* Manchester: Manchester University Press.

Benson, M. and O'Reilly, K. 2009a. Migration and the search for a better way of life: a critical exploration of lifestyle migration. *The Sociological Review*, 57(4), 608–25.

Benson, M. and O'Reilly, K. (eds) 2009b. *Lifestyle Migration: Expectations, Aspirations and Experiences.* Farnham: Ashgate.

Bourdieu, P. 1984. *Distinction: A Social Critique on the Judgement of Taste.* [R. Nice translation]. Cambridge, MA: Harvard University Press.

Bourdieu, P. 1986. The forms of capital, in *Handbook of Theory and Research for the Sociology of Education*, edited by J. Richardson. New York: Greenwood, 241–58.

Bourdieu, P. 1996. *The State Nobility: Elite Schools in the Field of Power.* Stanford, CA: Stanford University Press.

Buller, H. and Hoggart, K. 1994. *International Counterurbanization.* Aldershot: Avebury.

Cohen, A. 1982. Belonging: the experience of culture, in *Belonging*, edited by A. Cohen. Manchester: Manchester University Press, 1–17.

Fog Olwig, K. 2007. Privileged travelers? Migration narratives in families of middle-class Caribbean background, in *Going First Class? New Approaches to Privileged Travel and Movement*, edited by V. Amit. Oxford: Berghahn, 87–102.

Friedmann, J. 1994. *Cultural Identity and Global Process.* London: Sage.

Giddens, A. 1990. *The Consequences of Modernity.* Cambridge: Polity.

Giddens, A. 1991. *Modernity and Self-Identity.* Cambridge: Polity.

Goldring, L. 1998. The power of status in transnational social fields, in *Transnationalism from Below*, edited by M. Smith and L. Guarnizo. New Brunswick, NJ: Transaction Publishers, 165–95.

Gupta, A. and Ferguson, J. 1992. Beyond 'culture': space, identity and the politics of difference. *Cultural Anthropology*, 7(1), 6–23.

Harvey, D. 1989. *The Condition of Postmodernity: An Enquiry into the Origins of Cultural Change.* Oxford: Basil Blackwell.

Hastrup, K. and Fog Olwig, K. 1997. Introduction, in *Siting Culture: The Shifting Anthropological Object*, edited by in K. Hastrup and K. Fog Olwig. London: Routledge, 1–16.

Herzfeld, M. 1992. *The Social Problem of Indifference: Exploring the Symbolic Roots of Western Bureaucracy.* Chicago: University of Chicago Press.

Hoey, B. 2005. From pi to pie: moral narratives of noneconomic migration and starting over in the postindustrial Midwest. *Journal of Contemporary Ethnography*, 34(5), 586–624.

Inda, J. and Rosaldo, R. 2002. Introduction: a world in motion, in *The Anthopology of Globalization: A Reader*, edited by J. Inda and R. Rosaldo. Oxford: Blackwell, 1–34.

Ingold, T. 1993. The temporality of landscape. *World Archaeology*, 25(2), 152–74.

Ingold, T. 1995. Building, dwelling, living: how animals and people make themselves at home in the world, in *Shifting Contexts: Transformations in Anthropological Knowledge*, edited by M. Strathern. London and New York: Routledge, 57–80.

INSEE, n.d. Populations légales 2007 pour les départements et les collectivités d'outre-mer. [Online]. Available at: http://www.insee.fr/fr/ppp/bases-de-donnees/recensement/populations-legales/france-departements.asp?annee=2007 [accessed: 21st January 2010].

Kearney, M. 1995. The local and the global: the anthropology of globalization and transnationalism. *Annual Review of Anthropology*, 24, 547–65.

Levitt, P. 2003. Keeping feet in both worlds: transnational practices and immigrants incorporation, in *Toward Assimilation and Citizenship: Immigrants in Liberal Nation-States*, edited by C. Joppke and E. Morawska. London: Palgrave, 177–94.

Levitt, P. and Glick Schiller, N. 2004. Conceptualizing simultaneity: a transnational social field perspective on society. *International Migration Review*, 38(3), 1002–39.

Lewellen, T. 2002. *Political Anthropology: An Introduction.* Westport, CT: Greenwood Publishing Group.

Massey, D. 1994. *Space, Place and Gender.* Cambridge: Polity Press.

Morawska, E. 2003. Immigrant transnationalism and assimilation in the United States: a variety of combinations and the analytic strategy it suggests, in *Toward Assimilation and Citizenship: Immigrants in Liberal Nation-States*, edited by C. Joppke and E. Morawska. London: Palgrave, 133–76.

O'Reilly, K. 2000. *The British on the Costa del Sol: Transnational Identities and Local Communities.* London: Routledge.

O'Reilly, K. and Benson, M. 2009. Lifestyle migration: escaping to the good life, in *Lifestyle Migration: Expectations, Aspirations and Experiences*, edited by M. Benson and K. O'Reilly. Farnham: Ashgate, 1–13.

Oliver, C. 2008. *Retirement Migration: Paradoxes of Ageing.* London: Routledge.

Rapport, N. 1993. *Diverse World-Views in an English Village.* Edinburgh: Edinburgh University Press.

Reay, D. 2008. Psychosocial aspects of white middle-class identities: desiring and defending against the class and ethnic 'Other' in urban multi-ethnic schooling. *Sociology*, 42(6), 1072–88.

Rodríguez, V., Fernández-Mayoralas, G. and Rojo, F. 1998. European retirees on the Costa del Sol: a cross-national comparison. *International Journal of Population Geography*, 4(2), 91–111.

Rozenberg, D. 1995. International tourism and utopia: the Balearic Islands, in *International Tourism: Identity and Change*, edited by M. Lanfant, J. Allcock, and E. Bruner. London: SAGE, 159–76.

Savage, M., Barlow, J., Dickens, P. and Fielding, T. 1992. *Property, Bureaucracy and Culture.* London: Routledge.

Theodossopoulos, D. 2003. *Troubles with Turtles.* Oxford: Berghahn.

Torresan, A. 2007. How privileged are they? Middle-class Brazilian immigrants in Lisbon, in *Going First Class? New Approaches to Privileged Travel and Movement*, edited by V. Amit. Oxford: Berghahn, 103–25.

Williams, R. 1973. *The Country and the City.* London: Paladin.

Chapter 5
Ambiguous Foreigners: Neighbours Share More Than Geographical Space

Jacqueline Waldren

Resident foreigners own the largest properties in Deia, a village of 700 inhabitants on the northwest coast of Mallorca, Spain and draw on assumed superior knowledge, wealth and cosmopolitan experience to influence and appropriate what they deem as village landscapes and activities, often in contrast to Mallorquins' experiences and expectations of village life. Foreigners have appropriated local and space-specific symbols (landscape, climate, architecture, celebrations, public spaces, cafés, food styles and more) and now interpret and share these with other foreigners. The diverse and expanding foreign population claims to represent 'local people and places' and it is their images that are projected in conversations and publicised in foreign press. Musician Andrew Lloyd-Webber's favourite restaurants are mentioned in *The London Times*, hotel hideaway La Residencia (once owned by Richard Branson, now part of the Orient Express group) is featured as the 'place to be seen' and Robert Graves' poetry and prose, written over his 50-year residence in Mallorca, are on display in his home-turned-museum. In 2009 *Forbes* magazine listed Deia as the tenth best place to live in Europe.

How has this village, described by Catalan artist and writer Rusiñol in his book *The Island of Calm* (1929) as a 'still life waiting to be painted', become a rural idyll and a rendezvous point for international jet-setters, nouveau riche, travellers seeking luxury and 'authenticity' and tourists following guidebooks from beach to village to mass resort? What is it like for a local (whatever that means today) to live in the midst of this kaleidoscope of colourful visitors, amazing landscapes, nature, music, art, fancy restaurants, boutiques, real estate offices and traffic? How have those who arrived from abroad during the past three decades become the spokesmen and women, entrepreneurs, developers, shop owners and representatives of the village? What has happened to the local family connections that once provided powerful decision makers and authoritarian leaders?

The village council is still made up of locals, young men and women whose families have lived in Deia for generations but who often have little political experience. Politics are no longer just local; city politicians support and orchestrate much of what goes on in the village, so village council members are guided according to party directives. Many of the formal local roles within the village are backed by the informal power of foreign investors. Development has been financed by outsiders since the 1960s, and today there is competition among new money and long-term residents' ideas, new power brokers and old money, and inside networks (both local and foreign) that operate secretly until projects are finalised, allowing no room for criticism.

The village church is more cosmopolitan than the council, and recognises the varied nationalities of its congregation. Mass and holiday celebrations are introduced in German, English and Spanish, and then the services are carried out in Catalan (of which the local spoken Mallorquin is a dialectal variant). At Saturday mass the congregation is half foreign and half local, and on Christmas Eve, Three Kings and Easter it is 60 per cent foreign and 40 per cent local. Only on 24 June each year, which is the village saint's day for St John, or at a funeral do the locals' extended families reunite from their various homes around the island. Their return is a reconnection to their unique place of birth, its landscape, kin and memories of a different past. Although they share acquaintances with many long-term foreign residents and their children, much else they observe seems quite changed. After years of working in other parts of the island, they return to their birthplace more as tourists than as locals.

When I, a 30-year resident, walk down the main street I see the supermarket is owned by a local family, but the majority of its goods are imported and aimed at foreign tastes: chutneys, pâtés, French and Spanish cheeses, fruit from South America. The newsagents across the street (owned by the same local family) sells *Der Spiegel*, *Le Monde*, *The Daily Telegraph*, *The Guardian*, *The Independent* and *The Times* as well as journals, books, newspapers and magazines in English, Spanish and German. The city-built cultural centre *Sa Tanca*, where artists can exhibit their paintings, sculptures and artefacts of wood, clay or silk attracts mostly foreign artists, while the few local artists prefer to show their work in commercial galleries in Mallorca's main city, Palma. This centre is also used communally for carnival celebrations and other secular events such as neighbourhood association meetings, special lectures, exercise classes and political meetings prior to national or local elections. Public space on the village's main street, a road that connects it to Valldemossa eight km to the south and to Soller eight km to the north, is dominated by foreign-owned restaurants, cafés and boutiques selling Burberry-style classic clothes, designer fashion and Indonesian, Indian, Mexican and Asian quality goods. Two German real estate offices are the first sight one sees on reaching the village high street. The only all-Mallorquin space I see is a small room under the cultural centre, next to the doctor's office (the Galician doctor of the past 38 years has just retired and been replaced by an Argentinean), where the 'Third Age' (those over the age of 55) meet a few times each week to play bingo, hold meetings or plan excursions. A tiny nondescript green shuttered door (like those on most of the village houses) opens into 'their' space. The walls are covered with landscape paintings donated by local artists. I recognise this anomaly, but visitors see only the quaintness of the stone buildings up and down the streets and the beauty of the terraced landscape. Tourists feel welcomed and are understood when they order 'white coffee' or 'American coffee' in the cafés, as well as 'cake' or a 'sandwich'.

In this culture, once described as 'macho', it is the local women who have become small business owners and enjoy the benefits of college educations, keep-fit classes, gymnastics, Pilates and yoga. Women outlive men, while the old men express disappointment in their sons, who choose odd jobs such as gardening, house painting, doing repairs, construction or working as waiters, spend too much time in the cafés, marry late (or not at all) and prefer cars, boats and motorbikes to family life.

One must ask: has there been a spatial takeover or has prosperity, lifestyle, family and generational changes, individuality and education fused the vast differences once so clear between peasant culture and foreign visitors and residents? Do foreign residents appropriate local or place-specific symbols expressing ownership and entitlement, or is this part of a shared development that is the outcome of over 50 years of foreigners settling into the village? Who are 'the locals' now that three generations of foreigners have lived and died in Deia?

Space and Identity

I would like to rethink the concepts of 'insiders and outsiders', discussed in my 1996 book of that name, by focusing on the appropriation of space, place and practises of localising by foreign residents. 'Outsiders' is a category which can only be identified by its opposite, 'insiders'. However, after a hundred years of coexistence, the division is not so clear; foreigners have been integrated into village life, are more visible on the streets and public areas of the village, and foreign names appear on the local birth register and on tombstones in the cemetery. It has become difficult to identify locals from foreigners.

Through the process of planning an exhibition, I discovered that the spatial 'realities' of a locale are social, with different physicalities experienced by individuals. Through the arrangements of photographs, paintings, poetry and prose by both local and foreign artists, I observed that while foreign artists gave themselves a new 'local' identity by forming groups such as *Es Deu Des Teix*, named after the mountain ranges that surround the village, local artists identified their work with that of world-famous artists who worked in the village for as little as two weeks. Local Spanish and Mallorquin writers also proclaimed their affinity to a wider world of Spanish literature by associating their work with famous Hispanic authors who have visited the village (in some instances for as little as a few days), including Gabriel Garcia Marquez, Jose Luis Borges and Julio Cortázar. The concepts of 'space', 'place' and 'identity' fluctuate according to the circumstances. While local artists portrayed ancient and modern landscapes where every tree, river, stream, mountain and house had a name and form captured in classical landscape painting, the art produced by foreigners disconnected and reformed the landscape and people, creating abstractions open to multiple connotations.

Space and identity are constituted inside relationships and the use of terms like 'insiders' and 'outsiders' come about on different levels and in different circumstances, and may not be mutually meaningful. 'Us' and 'them' concepts were not based on larger societal distinctions, but drawn into the more immediate perspective of 'local community', while a 'global perspective' that goes beyond the here and now gave a wider connotation and allowed simultaneous claims to local connectedness and global mobility. Relationships are formed over time where people meet, where music or art is displayed, on the street, in the cafés, shops or at home. How one is greeted or acknowledged reveals these relationships (Waldren 1997: 23). Those of us both local and foreign who have lived the changes in Deia have memories of 'the way it was'. Equally the young, born and bred here, have internalised social knowledge and values, on-going family and social connections and new values and visions. All can make a claim to space and place, but have different priorities. Each feels it is their own, and the insider/outsider dichotomy seems less marked than when I wrote my book *Insiders and Outsiders* in 1996. On the surface, publicly there appears to be social and spatial reversal today: 'outsiders' (foreigners) are in-sight (seen more on streets, in cafés and so on) and 'insiders' are less and less visible. Yet both are essential for the infrastructure and consumption of services, commerce and bureaucracy, governance, building permits, upkeep of land and gardens and the ethos so essential to the ideal of a bohemian, peasant 'village' that attracts its admirers. At times, Deia seems a tightly formed community; however, at others it has no physical, social or artistic borders that confine people to place, as they describe links to world events, social, artistic or environmental concerns or other villages, towns, countries, family members, and friends in far away places.

In discussing conceptions of space (Waldren 1996: 34–58), I observed that 'descriptions and interpretations will differ markedly according to the age, gender, nationality, experience, work and interests of the person being interviewed. The ways that people order, value and express (often as much through actions as in words) their conceptions of space and knowledge of their "worlds" differs' (Waldren 1996: 40). Foreigners seem to share a concept of timelessness and a landscape devoid of people (other than themselves and other foreigners). In contrast, locals hold ambiguous feelings about 'their landscape'; they appreciate its beauty and their history and family connections to the land, but know how much tiring work was required to produce their livelihoods in the past.

In anthropology, 'landscape' has been used to refer to the meaning imputed by local people to their cultural and physical surroundings – how a particular landscape looks and feels to its inhabitants. However, the landscape in Deia seems to mean different things to the diverse population of this cosmopolitan village. To some, it is not just land, but families, work, social and religious associations, history, time, space and location carried in the mind and mediated in experience. For others, it is leisure, beauty, myth, pristine nature and social encounters with like-minded people. For all today, the landscape is redolent with past actions, history, ancestors and myths which are used differently by each in defining social groups, all of whom inhabit their various creations of 'the village'.

The concept of a lived space is phenomenological, emphasising individual experiences of the world. How do social groups interact with landscapes that are partly structured by previous groups? The social landscape is both content and context, enacted and material. Foreigners have claimed public spaces as 'theirs' by using them for extended periods of time, which interferes with local rhythms of work and leisure. The contests and tensions between different actors and interests in the construction of space can be explored. Cafés where 'local' men once sat playing cards for hours after work now choose to discourage these practises in light of increased foreign and tourist customers. I realised that the public spaces that formed part of the locals' concept of the village had a separate meaning for foreigners and business owners. Commercial priorities replaced long-term social habits. The restricted use of these physical spaces by locals altered the communication and action within these public social spaces. Local men and foreign residents had far less contact with one another and foreigners/tourists gained priority in these spaces. Concepts of identity and space are formed and reformed over time according to the social, political, economic and ecological developments within and beyond the village.

'The village' has expanded borders in the minds of the locals. In the first half of the twentieth century, young women who came from villages and towns in the centre of the island to pick olives, were included in social events and returned after a few years of courtship to marry local men. Their family and kin became part of an extended village socially, geographically and economically. Soller, a town eight km to the north, has been a trading and commercial centre for Deia. The marriages that ensued from these connections extended the sense of 'community' beyond the village borders. People in many different parts of the island trace family members, occupations and histories to common ancestors. Inter-village marriages and exchange networks were common and continue to form social and spatial links.

It is time to recognise that place, like voices, can be local and multiple. For each inhabitant, a place has a unique reality, one in which meaning is shared with other people in different places. The links in these chains of experienced places are forged of culture and history (Rodman 1992: 643). 'Rather than places becoming exemplars of our concepts they should be seen as, to varying degrees, socially constructed products of other's experiences' (Rodman 1992: 644). Tourists and foreign residents view the village from various perspectives: romantic, idealistic and spiritual, and liken it to a paradise that reflects timeless permanence. Drawing on Tomas, Sheppard and Walter's assertion that 'landscape is never value-free, and can function as a force in the formation of ideology' (2001: 542), we see the expatriates' creation of Deia as a mythical locality, an embodied experience of the world in which there is no Cartesian split between mind and body, culture and nature. The 'non-conventional' becomes 'conventional' for some people. 'The socializing role of bodily experience forms an ideology made material through landscape and may serve to naturalize it in such a manner that it may be impossible to imagine a world that is different. Movement, memory and daily routine within a landscape may work to create particular senses of personhood as well as ideas about appropriate action in different contexts' (Tomas, Sheppard and Walter 2001: 547).

The relationship between place and space and between insiders and outsiders' images and representations are dependent on cultural and historical context. Global and local perceptions and social relations have formed the place we call Deia today. According to Bender (1993: 1), 'Landscapes are created by people through their experience and engagement with the world around them'. Deia's landscape reveals the input of Romans, Moors, Catalans and today's Mallorquins, international residents and tourists. These people have engaged with it, re-worked it, re-imagined it, appropriated it and altered it. The scene one finds today is quite different than in the past. Houses have been built up the mountainsides and down by the cove. Both were once protected by laws prohibiting building in those areas. However, laws can be altered, and as the once profitable olive and citrus growing activities decreased, the lands became highly valued for urban development. These areas of 'new houses' were sold to or built primarily by foreigners. That meant new neighbourhoods were formed by people who knew little about the workings of the village and the social and physical realities of life in Deia.

According to Tilley (1993), landscapes are socialised through material human activity, such as the creation of dwellings and roads, just as physical features of the landscape, such as trees, mountains and oceans, shape human action. Every inch of Deia is privately owned, often named after its original owners, characteristic landmarks, elements of the area or activities carried out there. These names are metaphors of social relations both past and present, and reveal the social, economic and cultural history of the village (Waldren 1996: 78). The rural idyll sought by foreigners had little to do with this local history. However, unencumbered by tradition or social pressures, they had an important input into the improvement of local infrastructures including changing building laws, privatising water sources and sewage and rubbish collection.

Fiction, Fantasy and Nature

Much of what was written about Deia in the twentieth century was fiction, fantasy or tales of magnetic forces emanating from the mountains that inspired poetic expression (see Graves 1965, Alegria 1998, O'Brien 1988). As a resident and anthropologist, I wanted to discover the 'realities' behind the myths. What relationships did these foreign authors develop with the place and the locals? How did they perceive the village and its people? Were 'we' resident foreigners different from those referred to as tourists, were we insiders or outsiders or did we have changing identities in diverse circumstances? I have written about these issues in various publications (Waldren 1996, 1997, 2002, 2007, 2009).

Between 1975 and 1980 the foreign contingent continued to alternate between a small number of artists and writers committed to their individual pursuits and seasonal visitors. Some occasionally coalesced to deal with a local development that infringed on their privacy or threatened the beauty of the village. As the numbers and variety of foreigners increased, the ritual sharing between locals and foreigners decreased. Locals began to put a price on services and exchanges which were once viewed as reciprocal and their contact with new foreigners was mainly through commercial and financial transactions. The kind neighbour who once offered a home-grown cauliflower or freshly picked oranges now had foreign neighbours all around and felt she/he could put a price on these goods, or just leave them to fall on the ground and be used for compost. New arrivals who did not speak Spanish became reliant on long term foreigners for information and mediation of life in Deia. The public arena where casual encounters had occurred with locals and foreigners was now overcrowded.

The first foreign owned café-bar was opened in the 1970s. It attracted mostly foreigners and some locals who were beginning to participate in foreign public social life. Young local single men saw this as a means of getting closer to foreign women, who seemed so much 'freer' than local girls (this was the time of Franco/Catholicism). There was popular music, beer and wine, chess games going on in the corner, people milling about drinking, smoking, laughing, flirting or dancing. However, some long-term resident foreigners began to spend less time in cafés. Those who had arrived in the 1950s and 1960s as young men and women and found a few others with whom they could share their youthful hopes, dreams and aspirations had grown older together in Deia. By the 1980s, they left the regular café scene, with its heavy drinking and wild parties, to concentrate on their work, the families they had formed and the maintenance of the lifestyles they had created. They had acquired property when the prices were very low and had established permanent homes from which to pursue their dreams of art and poetry.

To the locals, the withdrawal of the foreigners with whom they were most familiar from public spaces seemed like a rejection. Social boundaries were fluid within the cafés, where everyone mixed freely. People did not casually visit one another's homes and boundaries became evident as people were less accessible. If they did not live close to one another, contact was limited. Long-established modes of communication and adaptation between foreigners and locals began to break down, and an increased awareness of symbolic boundaries occurred. Unintentionally, the long-term foreigners became associated with the newer foreigners as locals had little or no contact with them. Meetings in public social spaces seldom occurred, and when they did it was as though one had been away and had to catch up or reminisce about the past they had shared. The locals that did frequent the café seemed to retreat to the bar stools and stood around the bar talking among themselves, while the foreigners occupied the tables.

This separation of long-term residents from public spaces served both to control access to them and to develop a greater interest in their lives by locals and new foreigners. Stories about them become part of Deia folklore to be recited to newcomers in the café. Separation adds mystique, but it also calls attention to inequalities between locals and foreigners which had been less obvious in the past. Unlike the Mallorquin landed gentry, the long-term residents had idealised the local peasant life, extended families, beliefs and traditions. Economic differences were known but not highlighted either between foreigners or with locals. Almost everyone presented similar public images in old sweaters and jeans and voiced the same concerns when the price of cognac or beer went up from five to six pesetas. Mutual respect among local 'peasants' and artists and writers with paint smudged trousers or ink stained fingers was evident as they patted one another on the shoulder and loudly exchanged greetings in Spanish.

Most of these long-term foreigners have contributed to the preservation of Deia's past. Some tried to control the negative effects of time and modernity in the village, or at least in their own lives. Despite their efforts, the materialism of Western societies had begun to creep into Deia life and the idyllic peasant is a figure of the past. Changing local values, increased wages and mobility plus the arrival of different kinds of foreigners began to alter their paradise. Rather than the inside world dominating their existence, the outside world was now attractive, accessible and desirous to local Deianencs and to the children of foreigners who had grown up there.

Increased numbers of foreigners bought houses for holiday use only. With café life only one of the foci of social interaction, new homeowners and especially summer-only residents might reside in the village without ever getting to know either locals or long-term residents. Home entertainment, dinner parties, DVDs and television have replaced much weekday café time and it is mostly the new home owners and visitors who are able to afford the increased cost of dining and socialising in public spaces on a regular basis. The restaurants are fully booked by these new residents and hotel guests. Locals and residents might only be seen out on a Saturday night or at a fiesta and are generally astounded by the numbers that fill the seats and aisles when live music is performed in the local *Sa Fonda* café. Some long-term residents and locals greet one another; however, few others can distinguish between new residents, locals or tourists. Deia has always attracted a variety of people, and there are few places in the world where one can be surrounded by such an assortment of politicians, theatre and motion picture actors, artists, writers, shop assistants, farmers, office workers, tourists and locals, all gathered together on a terrace, listening to music under the stars, idly chatting or intensely conversing and taking no notice of one another except as social beings in a special space. These multiple attractions and unique qualities of Deia are the fictions, fantasies and realities that novels and the media promote far more often than they do for most other travel destinations.

Accommodating Change

Since the 1980s increased economic opportunities brought in by tourism have lead to a concerted effort by village authorities to change from within, to draw on local potential and develop local infrastructures and environment to accommodate residents and tourists without losing the beauty and ethos of Deia. The humanly shaped and preserved landscape in Deia began to show signs of neglect when tourism had drawn many local people into the rapidly developing services industry. Some men left the village to find work in the developing tourist areas on the coast, creating problems for older locals who no longer had the energy to keep areas cleared, terrace walls maintained, soil turned or to harvest the olives and citrus fruit. Changes in family, social life and gender identities were also evident. Young men no longer lived at home until they married. Girls went on to university rather than marrying young like their mothers; cohabiting or pre-marital sex were condoned. Building laws were drawn up and enforced, protected rural areas were identified and exempt from development. However, developers offering vast sums of money to the City Hall began to counteract these protective laws, and now buildings that extend beyond the church, around the coast and high into the mountains have altered the landscape and the concept of 'the village' considerably.

Paths once open to walkers who carefully pulled up a twisted wire holding a gate of wooden slats and replaced it as soon as they passed through are now cut off and marked 'PRIVATE'. Many foreign-owned houses have been surrounded by fences and heavy iron gates to close off the properties. Ownership, hidden lands and lives seem to be marked out by these enclosures. Behind many of these fences rolling lawns have been planted. Lawn is not natural to Deia with its stone and dirt earth, summer heat and water limits. It is a luxury which involves great expense and yet it is sought after in these new homes. Water is a commodity in Deia. Its consumption is measured and residents pay the highest rates for water on the island. There are shortages in the summer, and an over-abundance in winter when water rushes in torrents to the sea.

The climate, earth, growing seasons and the area's conditions of wind, rain, occasional freezes, extreme temperatures, sea air and water supply are all unimportant to these new owners who come mostly in the sunny summer months and employ gardeners and caretakers to look after things in their absence. Multi-million euro houses have been built on the mountainside and down by the sea, and the buyers brought with them ideas of landscapes from their imaginations, magazines and urban living. The symbiosis of nature and culture that has so long been a quality of life in Deia has not entered their consciousness and the affinity local people have with an area of land, its topography, waters, rocks, trees, paths and boundaries is rapidly declining.

Government subsidies have aided the city council in marking out paths and trails as nature treks for hikers in the hopes that island awareness of ecological issues will attract people to come to Deia seeking nature's bounty. The landscape has been commoditised and as locals become aware of the quality of their landscape and architecture the environment is transformed to meet the desires of 'others' (the wealthy, foreign and tourist). As Hirsch and O'Hanlon (1995: 65) point out, social space is not uniform and is constituted by different densities of human experience, attachment and involvement. It is socially produced and groups and individuals act out their lives in different spaces, in different ways. They also re-create the places and spaces to adapt to changing values over time.

The landscape manifests all of these changes. The myth of paradise remains an important factor in the marketing of destinations. In Deia, locals and guests have created myths about each other which give them a sense of belonging, and by re-telling these stories of the past to new visitors they are endeavouring to maintain the illusion of a timeless, tradition-based setting. The rare fascination the landscape has exercised over the imagination of generations of outsiders and insiders has paved the way into the future for both. However, so much that is now developing is based on urban images of the countryside and attempting to control nature in a way that may become detrimental in the future.

Deia is home to a resilient people and place that have adapted to multiple changes to lifestyles and landscapes. The cultural production of tourism in Deia seems to be a combination of foreign visions, tourist expectations and local awareness of Deia as a product to be consumed. Perhaps the practices, visions and ideologies of the foreign residents have begun to transform Mallorquins. 'The village' has been absorbed by the outsiders: expatriate residents, seasonal home owners, tourists in hotels and rental properties and day visitors in rented cars or hikers on the local coach service from Palma. Public services, restaurants and cafés are predominately foreign-owned: eight of the 11 restaurants are owned by foreigners, the two large hotels are owned by foreign companies and some long-term foreign owned properties are available to rent in the summer months when the resident owners prefer to be away from the heat and busy life in Deia and have come to rely on the increased rents to subsidise their pensions or limited incomes. Cafés and restaurants vie for business with reduced price lunch menus that feature Italian, French and cosmopolitan food or Spanish 'tapas'. Three cafés are owned by locals and, like the restaurants and hotels, are patronised mostly by foreigners. All of these businesses rely on family members or seasonal foreign employees (from Europe and Latin America) as young Mallorquins are no longer willing to work for others in the service industries. Those that do still work in cafés have mastered English, German and French through their interactions with customers and enjoy the cosmopolitan ambience of their work and the social life that it leads to.

Does all of the above support the early assumptions that tourism will destroy 'local' cultures? Talking to local people and long-term residents, it becomes clear that their diverse perspectives on village life make this a question that must be answered by individuals rather than making generalisations. The island council promotes local and national celebrations that re-affirm Mallorquin identity through language and history. Local councils organise village saint's days, traditional activities are made social: the annual sacrifice of a pig and the making of sausages is now a public social event, carnival is celebrated by the entire village dressing in costumes with cash prizes given for the most original, and a celebration of the 'Heydays of Creativity in Deia 1960–1980' included art, photography, film exhibitions, poetry readings and concerts. The photographs portrayed school classes, beach parties, café scenes and local events all peopled by the locals and resident foreigners who formed the Deia which visitors find today. All of these events bring together locals and varied foreigners in a constant interplay of identities and localities. Certain values from the past are re-interpreted and transposed and acquire new meaning as people put them to use in structuring their environment and social relationships with others.

Contrary to concepts of cultural demise, for most locals and those of us who have lived the changes, it is fascinating to see the impact Deia's landscape, myths, ethos, history and characters have on visitors. We have lived through the episodes and processes of change that are not reflected in what is discovered by today's visitors. We know the idyllic village in the mountains is not what it used to be, but for many people this is not a problem (the unreality of seeming idyllic was more difficult). The foreigner no longer represents the unknown, outside world. The locals have reached what they believe is equal ground – they too have access to the outside world, knowledge and incomes. There are no social or economic boundaries that make them feel different from anyone who visits their village and it is their expertise as builders, gardeners, electricians and plumbers that are highly valued by the newcomers. What seemed at the beginning of this chapter as an undesired appropriation of place and space by foreigners has actually created an ambience of familiarity for visitors and a safety net for locals who no longer have to compromise their identities to please or serve outsiders and have gained a respected personal identity as part of the landscape they understand, which the generations before them created. The foreign residents are seen to help maintain the businesses that underpin the performance of village life.

As the twenty-first century world experiences economic crisis, one wonders how a small village like Deia can survive. Tourism and foreign visitors are down by 30 per cent. The hotels close for one to three months each winter, restaurants may open only on weekends in the winter or remain closed for months at a time. However this can also be seen as a luxury, allowing for extended holidays and pleasurable home and garden pursuits. Two groceries, a bakery and two cafés provide the staples for daily life, and access to the supermarkets of the cities are just eight or 20 km away. The periphery of the village no longer encloses its inhabitants. Mobility via cars and regular coaches to the city, as well as internet and television expand the spaces and horizons once experienced as enclosed. Property is not selling as in the past, but the services needed to maintain the existing properties continue to provide incomes for many: state pensions support the elderly, and the young who usually have full employment in the tourist season collect their unemployment, travel and return to do odd jobs around the village in the winter before returning to work.

The long term residents and locals remember simpler living conditions and welcome the return to a less commercialised existence. However, there are many 'SE VENDE – FOR SALE' signs on houses and businesses, something seldom seen in the past when the only way one learned about property for sale was by word of mouth. With four realtors vying for buyers and foreign owners feeling over-whelmed by large mortgages and no potential buyers, some are feeling the 'crunch'. Service providers are also collecting payment directly on finishing a job rather than waiting, sometimes for months, to present their bill. These are all evidence of the global impacts on local lives.

Islands have often proved to be exceptions to most rules, and it is possible to imagine that anyone seeking an alternative lifestyle or the 'island of the mind' as discussed by J. Waldren (2002) might well look at an island like Mallorca. Deia has met the varied requirements of those seeking a 'different way of life' during the past century. Spain was especially cheap in the 1950s and 1960s, despite its oppressive government under the leadership of General Franco. As he encouraged mass tourism, he took no interest in the alternative lifestyles being created in the Balearic Islands (Mallorca, Ibiza or Formentera) where local attitudes offered an insular paradise in an inhospitable world. Locals were delighted to have foreign money coming into their bars, shops and restaurants, 'even if it came from beaded purses owned by long-haired aliens (hippies) of indeterminant sex' (Ferguson Thompson 1989: 178). These changing lifestyles constructed and de-constructed forms of national, regional, class and gender identity. Spain's historical development has been a response to the same dynamic processes of capitalist modernisation which defines the rest of the continent of Europe (Graham and Labanyi 1995: vi). The complex relations between the national and international which simultaneously erode and define Spain's cultural specificity cannot be ignored.

The lived practices and changing values of appropriation, identity and space in one small village have been demonstrated in this study of Deia. It suggests that under the guise of a 'traditional village people' there has been an acceptance of a fuller European or global identity after a long process of economic, political and social transformation. However, such a transformation means that certain values such as solidarity and sense of community must be seen as multifaceted and individualistic, and the prominence of consumer society cannot be denied. Still, the concepts of those 'inside' as Strathern showed (1981) was like the desert mirage, disappearing further into the distance as one approached. The symbolism of the boundaries was supposed to allow flexibility, as they could constantly be redrawn according to context, to be more or less inclusive. In Deia, locally perceived boundaries delineate separate entities (locals, newcomers, insiders, outsiders) under different headings in each new set of circumstances and community is not a thing with independent existence but, rather, a relational concept that is articulated differently through actions, discourse and symbols. The memories of various pasts, and the stories and myths that have evolved all combine to suggest that perhaps 'appropriation' is a misconceived approach to understanding multicultural lives as each group reinforces the others' sense of place. The foreigners' real and symbolic ownership, landscapes of the mind, fantasies and fictions combine with a diminishing local population's knowledge of real past struggles and social and political realities. As Edwin Ardener noted,

> That there is a multiplicity of identities that coexist from any single perspective is not strictly speaking a problem theoretically. It is one of the proofs – and one of the costs – of the apparent paradox of the continuity between the space and the individuals that constitute it. They are defined by the space and are nevertheless the defining consciousness of the space. (1989: 39)

References

Alegria, C. 1983. *Flowers from the Volcano*. Washington: George Washington University Press.

Ardener, E. 1989. Remote areas – some theoretical considerations, in *The Voice of Prophecy*, edited by M. Chapman. Oxford: Basil Blackwell, 211–23.

Bamford, S. 2002. On being 'natural' in the rainforest marketplace: science, capitalism and the commodification of biodiversity. *Social Analysis*, 46(1), 35–50.

Bender, B. 1993. *Landscape: Politics and Perspectives*. Oxford: Berg.

Cosgrove, D. and Daniels, S. 1994. *The Iconography of Landscape*. Cambridge: Cambridge University Press.

Ferguson Thompson, P. 1989. Ibiza, white island, in *Insight Guides: Mallorca & Ibiza, Menorca and Formentera*, edited by A. Eames. Singapore: APA Publications 175–85.

Graham, H. and Labanyi, J. 1995. *Spanish Cultural Studies: An Introduction*. Oxford: Oxford University Press.

Graves, R. and Hogarth, P. 1965. *Majorca Observed*. London: Cassell.

Hirsch, E. and O'Hanlon, M. 1995. *The Anthropology of Landscape*. Oxford: Clarendon Press.

O'Brien, E. 1988. *The High Road*. London: Weidenfeld and Nicolson.

Rodman, M. 1992. Empowering place: multilocality and multivocality. *American Anthropologist*, 3(94), 640–56.

Rusiñol, S. 1929. *Mallorca: The Island of Calm*. Palma: Imprenta Politecnica.

Strathern, M. 1981. *Kinship at the Core*. Cambridge: Cambridge University Press.

Tilley, C. 1994. *A Phenomenology of Landscape*. Oxford: Berg Publishers.

Tomas, T., Sheppard, P. and Walter, R. 2001. Landscape, violence and social bodies: ritualized architecture in a Solomon Islands society. *Journal of the Royal Anthropological Institute*, 7(3), 542–72.

Waldren, J. 1996. *Insiders and Outsiders*. Oxford: Berghahn Books.

Waldren, J. 1997. We're not tourists we live here, in *Tourists and Tourism: Identifying With People and Places*, edited by S. Abram, D. Macleod, and J. Waldren. Oxford: Berg Publishers, 57–74.

Waldren, J. 2002. Conceptions of the Mediterranean: islands of the mind, in *Islands in Prehistory*, edited by W.H. Waldren and J.A. Ensenyat. Oxford: British Archaeological Reports (BAR), 1–6.

Waldren, J. 2007. Reframing place. Time and experience: leisure and illusion in Mallorca, in *The Discipline of Leisure*, edited by S. Coleman and T. Kohn. Oxford: Berghahn Books, 73–90.

Waldren, J. 2009. Lifestyle afterthoughts, in *Lifestyle Migration: Expectations, Aspirations and Experiences*, edited by M. Benson and K. O'Reilly. Aldershot: Ashgate, 153–64.

PART III
Houses and Homes: Intimate Migrant Place

Chapter 6

A Reluctant Locality: The Politics of Place and Progress in Santo Domingo

Erin B. Taylor

When evoked as a counterpoint to globalisation, the term 'locality' conjures up images of communities where cultural traditions and social networks are closely integrated and rooted in Cartesian space. Localities are described as thriving as a result of global connectedness, and place-based sentiments and 'strategies of localisation'[1] are fuelled by collective desires to resist 'global localisation' in which economic capital, the state and technoscience produce locality in their own favour (Escobar 2001: 161). Apart from studies of refugees forced into new socio-spatial configurations by physical violence, most scholarly accounts suggest that strategies of localisation are effective political resources because they draw upon territorialised culture. This is especially true for indigenous movements, for which continuous physical and cultural occupation of land underwrites claims of ownership. It is also true for migrants reterritorialising their identities in new places, both in their residential communities and public spaces. The 'politics of recognition' (Taylor 1994) through which social groups seek to legitimate their claims to resources, social position and self-determination more often than not depend upon recognition of culture and society as rooted in place.

However, not all place-based political struggles can claim legitimacy based upon territorialised culture, and nor may they wish to. Squatter settlements share with refugee camps the ignominious status of being sites where place-making is a 'choice of the necessary' (Bourdieu 1984). This casts these settlements as illegitimate and their residents as 'vagabonds' who have migrated to poor urban communities 'because they have no other bearable choice' and who will move on when a better opportunity presents itself (Bauman 1998: 47). These migrants are made local – 'attached to a place, rooted, circumscribed, limited' – through their exclusion from domains controlled by state and capital (Asad 1993: 8). Yet residents often have little option but to invest in the production of locality – the development of a relatively bounded material environment and the objectification of social relations within it – with the result that, over many years, such communities become permanent, though they may remain marginal

1 Escobar refers primarily to the use of notions of place and territory by social movements in their political struggles, especially in response to the loss of local autonomy over place to control by states or transnational organisations.

and technically illegal. Their efforts to carve out a life in the city and achieve progress are strongly tied to the communities in which they reside. In the face of animosity by the state, the general public and economic capital, their efforts to construct housing and make place can be highly political. What happens to the politics of place when residents have formed a community reluctantly and their aspirations lie elsewhere?

In this chapter I present a case study of La Ciénaga, a squatter settlement in Santo Domingo, Dominican Republic, which is primarily populated by rural to urban migrants. I explore what is at stake, and who *has* a stake, in their politics of place. Locked into urban space, residents depend upon locality to resist their marginalisation and strive for socioeconomic mobility. They construct permanent dwellings, build familial and social networks and form institutions. Household efforts to make place are practiced alongside individual mobility strategies and collective efforts to politically integrate the community and modernise the material environment.

Nonetheless, continued material and symbolic marginalisation and concerns about social problems within the barrio fuel residents' aspirations to move from La Ciénaga to a 'better' suburb of the city or overseas. Efforts to generate political unity within the barrio are thwarted by these factors and by the residents' sense of their primary modes of belonging (place of origin, family and identification with Dominican culture and mainstream society) as being disconnected, or disconnect*able*, from the locality. Their claims are based upon being recognised as the *same* as outsiders, not different as is the case in Taylor's politics of recognition (Taylor 1994). Residents' production of locality is therefore often contradictory, simultaneously generating forms of social solidarity (in both formal and informal institutions) and creating social hierarchies (such as those based on race, ethnicity and point of origin). I demonstrate how, in a case where space 'is a key stake in the struggles of labouring people against the space-fixing practices of capital and the state' (Gregory 2007: 239), the politics of place draw more upon people's perceptions of their belonging outside of the locality and their aspirations to leave it, rather than their cultural and social embeddedness in the locality itself.

Localisation in the Dominican Republic

In August 2005 I was designing a survey in La Ciénaga and asked one of my neighbours for advice. He took issue with one particular question I had framed: 'Would you move away from La Ciénaga if you had the opportunity?' My neighbour argued that this question was redundant, because every single resident would answer 'yes'. If they answered 'no' they would be lying, because no sane person would choose to reside in La Ciénaga voluntarily. Indeed, when I collated the results of the survey, 96 per cent of the 300 residents surveyed reported that they would leave La Ciénaga if they had a choice. Who, they argued, would want to live in a *barrio marginado* (marginal barrio) characterised by delinquency, pollution and poor housing?

This result points to the radical localisation of barrio residents in Santo Domingo. The fact that residents of a squatter settlement wish to move is in itself unsurprising. What is notable, however, is how much research on Dominicans over the past decade has focused on their engagement with transnationalism, despite the fact that many Dominicans have very limited mobility. The Dominican economy depends upon remittances and tourism, and a long history of transnational relations has been a defining feature in the production of national culture. Notions of modernity and progress in particular have drawn upon Europe and the United States (Martínez-Vergne 2005, Hoffnung-Garskof 2008), and transnational migration has altered the way that Dominicans think about themselves and their nation (Grasmuck and Pessar 1991, Vicioso 2000, Ferguson 2003). Indeed, studies of Dominican international migration outweigh studies of Dominicans at home.

While an important object of study, this profusion of interest in *dominicanyork* and other transnational identities ignores the fact that the majority of Dominicans cannot migrate overseas due to the unfavourable position of the Dominican Republic in the global economy, and of the nation in the global moral order. In his ethnography of the beachside resort town Boca Chica and its poor neighbouring community of San Andrés, Steven Gregory (2007) explores the effects of globalisation on poor residents as they struggle to make a living. Most of these residents were first or second generation migrants from the countryside who came to Boca Chica to survive by serving wealthy tourists or producing goods for export. Their experiences of transnationalism were primarily of being locked into place while surrounded by movement of tourists and commodities.

Critiquing the tendency of globalisation discourse to focus on 'movement', Gregory writes about 'the lives, livelihoods, and struggles of people unable to move and "fixed" in space by economic hardship and by spatial practices restricting movement, citizenship rights, and access to a living wage' (2007: 6). Residents of San Andrés have daily contact with global flows through their work in tourism and export manufacturing, but they have little chance to achieve for themselves the mobility that they witness in others. Even if the residents of San Andrés were to save enough money from their meagre wages to buy a ticket to the United States, they would stand little chance of gaining even a visitor's visa due to their lack of plausible securities. They are incorporated asymmetrically into a 'spatial economy of difference', in which privilege is protected and hierarchy reinforced on a daily basis. In this hierarchy, the local loses out to the foreign, as white tourists enjoy the privileges of having money to spend and power to command. In contrast, residents' experiences and efforts to achieve *progreso* (progress) are very much bound to the locality in which they live. The positioning of poor localities close to city centres and tourist resorts reflects their class status as servants, poorly paid labourers and informal workers with no job security.

Place Politics I: Where Progress Resides

Thirty kilometers away, the majority of residents of Santo Domingo's squatter settlements do not make a living in transnational industries, yet their experiences of globalisation are similar. Approximately 300,000 people (11 per cent of Santo Domingo's population) reside in inner-city squatter settlements and barrios around the Ozama River (Fay and Wellenstein 2005: 97). La Ciénaga is one of the poorest of these barrios. It was first settled in the 1960s by rural migrants, and today it has approximately 18,000 residents who work primarily in the city's service industry as domestic servants, security guards, construction workers and street vendors.[2] Lack of employment and services combines with the stigma of residence in a poor barrio to restrict their socioeconomic mobility (Taylor 2009b). While 90 per cent of households own their home, none hold title to the land they occupy. Their houses are not worth enough to make it possible to sell and move to another community. As a result, most residents are long-term, and their children find it difficult to move out of the barrio.

Their emplacement, together with the repression of labour organisation and a history of authoritarian governments, has made grassroots politics in the Dominican Republic overwhelmingly place-based and focused on issues of residence and service provision. These politics are observable in the community's political organisations and in residents' daily practices. These include the construction of houses by residents, their efforts to gain land titles and campaigns for the material development of the barrio. Holston (1991) remarks how in Brazilian favelas 'autoconstruction', in which residents construct their own housing, provides residents with 'a new form of agency', but its political aims are unable to transform the structural position of the poor. Their localisation not only defines their immediate politics, but it also restricts the political field they can feasibly engage with. Gaining control of space gives some security, but only rarely geographic or socioeconomic mobility. It does, however, permit residents to carve out social spheres in which they feel a sense of belonging and autonomy. In Holstons' words, autoconstruction also permits residents to replace negative stereotypes of autoconstructed communities with 'new ones of competence and knowledge in the production and consumption of what modern society considers important' (1991: 462).

La Ciénaga is commonly referred to as *abajo el puente*[3] (under the bridge) because the main entrance is located under the Bosch Bridge, which links the

2 There are, of course, exceptions. Alternative professions include engineers, lawyers, secretaries, teachers, computer technicians and nuns, but on the whole, *cienigüeros* (residents of La Ciénaga) find that the range of occupations available to them is limited, so the barrio is relatively homogeneous in terms of income. Most incomes tended to range from 1,500 to 5,000 pesos per month (US$45–150). The highest salaries reported to me were 15,000 pesos per month (US$440), earned by a carpenter and a grocery store owner.

3 A local rendering of '*debajo del puente*'.

city on either side of the Ozama River. To outsiders, La Ciénaga contradicts notions of what a modern city – and the modern nation – should be. Historian Jesse Hoffnung-Garskof notes, 'because it was visible to anyone crossing the ... bridge, La Ciénaga became emblematic to Santo Domingo residents who commonly referred to any marginal settlement as "under the bridge"' (Hoffnung-Garskof 2008: 41). The material environment of the barrio – its visible poverty, predominance of shacks and small concrete houses, lack of services and narrow alleys – is interpreted by outsiders as a manifestation of collective moral disorder (Taylor 2009a). Bridges and highways have stood as symbols of progreso in the Dominican Republic at least since the American occupation of the 1920s. Looking from the bridge at La Ciénaga implies looking from a position of relative progreso to a site that is seen as its antithesis. The barrio is not just a local development problem: it symbolises the difficulties involved in realising national and individual prosperity and autonomy.

The exclusion of poor barrios from state visions of modernity has a long history. Dominican elites and intellectuals have viewed inner-city slums as impediments to progreso since at least the late nineteenth century. Teresíta Martínez-Vergne (2005) describes how the Dominican urban elite, comprised of intellectuals, politicians and a burgeoning middle class (Bosch 1992), were struggling to consolidate their power over the recently independent nation. 1865 marked the Dominican Republic's second attempt at independence, having briefly returned to Spanish rule in 1861 after its initial independence from Haiti in 1844. The political and intellectual elite shaped nationalism around notions of democracy and progress, calling for the implementation of 'modern agricultural techniques, secular education, and political participation as the cornerstones of the new nation' (Martínez-Vergne 2005: 1).

This idea of progress was based upon European liberalism and would ideally be inclusive of all Dominican citizens, but 'men of letters and men of state did not trust the capacity of the subalterns to make educated judgments about their own, let alone the country's, future' (Martínez-Vergne 2005: 50). They were particularly concerned with bringing the unruly countryside under the state's control, as it was dominated by *caudillo* (strongman) political leaders with their own private armies (see Wolf 1967), and was considered by urban elites to be a site of moral degeneration (Martínez-Vergne 2005). The large urban centres were fixed as the sites in which modern citizenship would be carved out. However, the urban working class presented a significant obstacle to the maintenance of order:

> For municipal authorities intent on improving the quality of city life, these populations represented serious obstacles to their goals. They circulated freely around the city, presumably cared little about their personal appearances and hygiene, exhibited 'scandalous' behavior, and thus strained the resources available to their superiors and public authorities to manage their existence. (Martínez-Vergne 2005: 127)

New municipal ordinances covered a range of social, economic and aesthetic ideals, including implementing construction codes, banning animals in the city, regulating dance halls and taxing stallholders in public markets. Social classes were spatially segregated, occupied different grades of dwellings and socialised in distinct venues. This policy was expanded during the dictatorship of Rafael Leónidas Trujillo, from 1930 to 1961. Taking advantage of the damage wrought by Hurricane Zanón in 1931, Trujillo cleared inner-city slums to make way for wide, Paris-style avenues and model suburbs complete with detached houses set back from the street to contain socialising to private spaces. Trujillo built bridges to expand the city and monuments to concretise the importance of past heroic events to the nation (Derby 2009). He mobilised state historians, government departments and the media to distance *dominicanidad* (Dominicanness) from Africa and orient it towards Europe. Trujillo's regime defined the national colour as *indio* (Indian) and disseminated a racist discourse of *antihaitianismo* (anti-Haitianism) to posit the nation as civilised and modern in opposition to Haiti's poverty and perceived primitivism. Migration from the countryside to the city was restricted, and public spaces were reserved for the light-skinned elite.

Despite continued government opposition, urban migration boomed after Trujillo's assassination. Trujillo had believed that rural people came to urban areas because they were 'bedazzled' by the modern promise that the city held. There was also little recognition under Joaquin Balaguer's subsequent presidencies (1966–78 and 1986–96) that Trujillo's policies had dispossessed thousands of peasants through the creation of large sugar estates held by United States companies in the early years of the regime, and later by Trujillo's own monopolistic practices. President Balaguer issued a series of decrees banning migrants from settling in the *zona norte* (north zone) around the Ozama River, evicted residents and refused to install services. The barrios continued to grow regardless.

Today, many barrio residents contrast the morality and healthiness of rural life with the corruption and pollution of the city, claiming that they would prefer to live in the countryside but it is an economic impossibility (Taylor 2009a). However, their aspirations for progress are unlikely to ever be realised within La Ciénaga. The barrio cannot be transformed into an ideal modern community while it remains overcrowded: indeed, an official name for a 'good' residential community is an *ensanche*, which means 'widening'. Small, overcrowded shacks are one of the major indicators of material underdevelopment. To attain the population density of an ensanche, approximately half of all current residents of La Ciénaga would have to leave the barrio and their houses be demolished. Collective politics of place are therefore concerned with movement out of the locality as much as residence within it.

Barrio residents' aspirations for the future include dreams of material wealth: the modernisation of the barrio, salaries that allow for the consumption of consumer goods and transnational lifestyles characterised by movement between diffuse places (Taylor 2009c). People's imaginings of place are always made with reference to the outside world; that is, with awareness of their place in the present social – and indeed, global – order:

> Sometimes I'd like to leave because when one sees things like Cyclone David one wants to live in the high part, but in many ways this is the best area because it's so close to the city, this La Ciénaga. But it's not a good barrio anymore. And at times it's my wish to leave here to live again in my village, in my place, in my village another time, in La Vega. One lives well there. But the church is here and I'm comfortable; the community in my village isn't like this one. This country *dominicana* is in darkness all over because the power plant isn't functioning. It's not like this in your country, is it? The light never fails there. (Samuel)

Samuel's comment reveals the tensions and ambiguities involved in the negotiation of place in modernity. His is a vision that looks inward to an idyllic past as well as outward beyond national borders. It is the city's poor that struggle most to resolve this dilemma, compelling them to be creative in developing their responses.

Their near universal placement of aspirations outside the locality notwithstanding, residents have made significant strides in constructing place within the confinement of the barrio. The vast majority of cooperative projects have been organised by, or are affiliated with, the local Catholic church and its Jesuit priests, who have an interest in forming a stable congregation to increase participation in church activities such as mass and prayer meetings. The construction of the barrio's first school, Virgen del Carmen, in 1987 was one of the first major communal actions achieved by the Catholic congregation.[4] It was organised by a nun called Maria Blanca and carried out in collaboration with the community, with men supplying labour and women carting materials to the site and preparing food. Every house donated a concrete block towards the building of the school and contributed labour according to their capabilities. Soledad, a middle-aged community activist, remembers that she had a bad leg and was ordered to rest by her doctor, but 'God gave me the strength to go on working'. The thought of God watching her and judging her kept her going back every day. Soledad feels that to struggle in daily life is to work towards God and a place in heaven. Similarly, Lidia remembers:

> I was involved when we, the parents, made the school that used to be here. We carried heavy things, we carried the food, we carried breakfast when the men were working. We helped a lot with the school. It was us. (Lidia)

4 The original school was subject to frequent flooding and has since been replaced by a four-floor composite school constructed by the government on higher ground.

Residents remember the construction of the school with pride. Community actions such as these stand as evidence of residents' civility and moral worth, in contrast to a normative discourse that speaks of the worthlessness of the urban poor. Furthermore, they foster a sense of collective identity and purpose among migrants who originated in different parts of the country. Resident relations have been institutionalised through the formation of legal community organisations such as CODECI (Coordination for the Development of La Ciénaga). CODECI is an umbrella organisation funded by the Catholic church that coordinates the activities of all the other community organisations. Its vision is to modernise the barrio so that existing residents can live there comfortably. The leader of the community's sports organisation explained:

> We want to *llevar mas adelante* (take ourselves forward). As they say, a diamond is made of carbon but once you polish it, it shines. That is, you bring out the value. As an institution we could prepare a fertile terrain, that is, the seed that we plant today will turn the barrio into fertile terrain for development. (Gabriel)

La Ciénaga has a number of organisations that work towards community development. These meet collectively every fortnight to discuss their activities and make joint plans. They carry out small projects individually, such as cleaning up stormwater drains, but most of their projects depend upon lobbying for funds from the municipal government, federal government and local and transnational NGOs. Recently funded projects have included mapping and naming all the alleyways, installing stormwater drains and building stairways and public paths. The Catholic church and the state jointly funded the relocation of some residents whose houses were dangerously close to the river's edge to apartment blocks in a suburb close by.

In recent years the community has achieved significant improvements to living standards, including better drainage and a more regular electricity supply. It has also managed to improve the barrio's public reputation through official media releases and press conferences (Taylor 2009b). In a survey I undertook of city (not barrio) residents, quite a few people expressed to me the belief that 'La Ciénaga is bad, but not as bad as before'. The material possession and transformation of space has resulted in its partial conversion to symbolic capital (Bourdieu 1984) in the form of material improvements and some (albeit limited) recognition of the community's legitimacy by the state and the general public. The presentation of the barrio as an organised urban community has been an integral part of this re-identification.

However, the church has difficulty unifying people to invest in a community in which they have little security of residence. The majority of residents involved in church activities are those whose homes are constructed in the 'good' parts of the barrio, which are least likely to be developed. The most important practice defining the barrio's politics of place is not collective political action, but rather household efforts to construct their own homes and transform them into modern concrete dwellings, within proximity of friends and family. This has been done in defiance of state evictions in 1977 and 1991, a military blockade of the barrio's entrance from 1991 to 1996 in which residents were prevented from bringing in building materials, and continuing symbolic marginalisation from the city's social life.

Place Politics II: The House as Locality

The production of locality in La Ciénaga – the development of the material environment and the objectification of social relations upon it – is more of a response to localisation than a desire to invest one's life in this particular place. Nevertheless, the community has provided residents with some control over their material and social environment. La Ciénaga exists precisely because the illegal nature of the settlement provided poor migrants with the opportunity to live in the city close to their family and friends, and to attain some measure of progreso by building a house and transforming it from a rural-style, wooden *ranchito* into a modern concrete house.

Every shack is constructed with the idea that it will be transformed into a house that would meet middle class standards of respectability. Over time, perhaps many years, residents improve their homes as resources permit, until they eventually conform to aesthetic (if not legal) norms of what a modern urban dwelling should be. Throughout the Dominican Republic, the ideal house is made of concrete block with a tiled concrete floor and a concrete slab roof, a model that provides much greater protection from hurricanes than the wood and tin shack. It has white metal shutters, bars on the windows and air conditioning in the main bedroom. As much as elsewhere in Santo Domingo, residents of the barrios aspire to this ideal, although it may take them decades to achieve or even be left for the next generation to complete.

Ironically, given that the barrio is widely perceived to be violent and dangerous, power over space allows the residents of Santo Domingo's barrios to maintain close-knit family units and a high level of community embeddedness that *capitaleños* (residents of the capital) say is disappearing in the city (see Taylor 2009b). Because the barrios were constructed by small family units on very low budgets, almost all housing in the barrios consists of small, individual family dwellings that are close to each other and to the street. The style of houses and their proximity create a social atmosphere. Since few residents can ever afford to leave La Ciénaga, social relations are surprisingly stable. Many people have lived alongside each other for 20 or 30 years, and family networks tend to be extensive. Often one person will migrate from the country, establish a basic dwelling, then bring family members to the barrio one by one. Their children may eventually construct their own homes in the backyard. As one resident explained to me:

> Our family relations are good because we have always shared our problems and happiness. We have always maintained contact. For a while I lived in another part of the barrio but I had to return. I missed being with them. My uncles also live close by, my aunt has always lived near us. Where one goes, the rest go and it has always been like this. We haven't been able to disperse ourselves. (Samuel)

The extended family depend upon each other's help on a daily basis, sharing food, knowledge and gossip. Over the years, Samuel's father Pedro and Pedro's cousin Altagracía have actively helped members of ther families establish themselves on the same street. As Steven Gregory (2007) has noted, the poor depend upon this social capital to overcome their economic marginalisation. One of Samuel's uncles commented to me:

> We are a very united family. If one of us is hungry and the other one can give him food we do so enthusiastically, and we always treat each other well. When we have some time we visit each other and greet each other with lots of affection. We're a family that is totally united because we were brought up to respect home life. This was the teaching that gave us such good unity. (Carlos)

For this extended family and many others, the benefits of constructing a house on land without a title far outweigh the alternative of struggling to pay the rent for a tiny flat elsewhere in the city. The area stigma that accompanies autoconstructed communities certainly impacts upon the ability of residents to find respectable and well paid employment, yet it also allows for greater independence and choice. It lets residents create their own social worlds that are seen as antithetical to the impersonal middle class, while engaging in consumption to express their right to a stake in progress and modernity. Residents struggle against localisation, but at the same time they appreciate, and indeed depend on, the sociality that their localisation brings.

Cienigüeros admire and aspire to the modern housing in middle class areas, but they perceive its life to be unsociable and boring. In particular, they criticise the middle class for living individualistic lives, emphasising the importance of social relations to counter the depersonalism of the global city. According to one of my neighbours, Felicia, 'There the people live well but they lock themselves up in their flats with bars. It's like living overseas!' Felicia's viewpoint is a common one, suggesting that progress is obtained at the expense of sociality. It is a view that is also espoused by the middle class about life overseas. A woman I lived with for a few months in Gazcue, a middle class suburb, had a Dominican boyfriend who lived in Miami. She had travelled there many times, but had no desire to sacrifice her Dominican sociality for the isolation of the United States. While barrio residents covet good quality housing situated in well-serviced, safe neighbourhoods, they are critical of the social distance inherent in the lives of the urban and wealthy.

However, as I demonstrate in the next section, residents' sociality is limited as they hold competing definitions of what progreso is, who should have access to it and who is capable of achieving it. They reproduce the hierarchies that emplace them within the community, developing a local moral geography (Harvey 2000) to classify different barrio residents and migrant groups based upon judgments of their material and symbolic worth. Perceptions of social divisions and problems within the barrio are a major contributor to people's reluctance to commit to place, but they may also guide the forms in which people engage with the creation of locality.

Place Politics III: Hierarchies of Belonging

Hoffnung-Garskof (2008) writes how, in the 1950s, the meaning of progreso in Santo Domingo shifted from social class, education and housing toward mass consumption. Previously conceived as a national strategy (albeit an exclusive one), from the 1950s onward it referred more to improving one's circumstances through constructing or renovating a home and filling it with modern material goods. This new mode of progreso was fuelled by an increasing orientation toward the transnational sphere, especially in response to imports of foreign goods and popular culture. But, as Daniel Miller (1994) has noted in relation to Trinidad, engagement with 'transience' through mass consumption entails a certain loss of respectability, which continues to be tied to the domestic sphere and formal institutions such as churches and schools.

Home ownership in La Ciénaga signals commitment to society, including one's neighbours, but also to a nationally endorsed modernity characterised by development and social stability. It requires commitment to place and, as Holston points out, it produces disciplined workers who depend upon their income to realise their life goals. Conversely, whereas home ownership and the forms of consumption associated with it are strategies to carve out security in the market, conspicuous consumption signals a loss of faith in the nation's future prospects and its possibilities for inclusion, instead marking a transient and transnational material culture as the place to realise aspirations.[5] The consumption of non-durable goods, instead of investing resources into the construction of a house, signals a rejection of stability and discipline, the breakdown of Dominican social fabric and its replacement with crime and corruption. Of course, this negative reading of consumption depends upon one's capacity to spend and the source of one's earnings. Throughout Santo Domingo, it is common to hear people speculate that the poor earn their money to consume through theft, or that they are neglecting their families and security by spending money on consumer goods rather than on improving their homes.

5 For a discussion of these issues in regards to Jamaican transnationalism, see Robotham 1998, Thomas 2004.

Certain groups of people in Santo Domingo are considered less capable of realising respectability than others, and in La Ciénaga this division is highly spatialised. The barrio has two definitive spatial sections: the high ground in the centre of the barrio (*arriba*), and the low ground that surrounds it (*abajo*). Arriba consists of Altagracía's sector (La Clarín), the main street leading to the barrio's entrance and the cliffs separating La Ciénaga from Guachupita. Most houses located arriba are built of concrete block and a large number of its original residents still live there, maintaining a community feel. La Clarín is the most desirable area of the barrio to live in because it does not flood and it houses the school, the Catholic church and CODECI's office. It also has the highest rate of church attendance out of the barrio's six sectors. In short, it offers domesticity and institutions of the sort that Peter Wilson (1973) associated with 'respectability' on the island of Providencia, Colombia.

Abajo is densely populated and its material underdevelopment advertises the poverty of its residents. Houses located abajo are small and primarily constructed of wood and tin. They are joined together by a network of alleys, which convey a sense of inaccessibility and alienation from the city. Few residents appear to know the area in its entirety: residents of the central region may not know their way around the riverbanks, and vice versa. Many people stick to the main paths through abajo to avoid getting lost in the more inaccessible, unpaved and dirty alleyways. A stormwater system was installed in 2005, but a lack of drainage continues to be a problem in some parts, where water muddies paths and human waste stagnates in open drains, providing a breeding ground for mosquitoes and disease. A government initiative has covered drains and concreted paths in much of the barrio, reducing illness and mud. The concrete paths also provide space for children to play between the close-set houses.

This division within the barrio corresponds with a city-wide spatial division in which all the barrios located abajo are criminalised. The reproduction of this spatial hierarchy within La Ciénaga can be seen as an attempt to differentiate the self from stereotypes of the barrio as being entirely criminal. Economic insecurity and overcrowding and the social changes wrought by them have resulted in a heightened sense of danger in the city and the barrio. Under such circumstances, there has been an enhanced focus on a pre-existing process of separating respectable and non-respectable worlds. The barrio is classified into high versus low space (arriba/abajo), but also private versus public space (house/street) and reputable versus disreputable architectures (churches and other institutions/bars and gambling places). It mirrors a city-wide moral geography that demonises the entire barrio through the association of poverty with social disorder.

Residents of La Clarín engage with this moral geography in a certain way: they resist the barrio's criminalisation by positing themselves as respectable, moral beings, while simultaneously verifying the association of poverty with social disorder through the division of space and people within the barrio. Unlike arriba, which has changed dramatically over time and now resembles a poor but relatively modern neighbourhood, abajo's underdevelopment and pollution challenges the barrio's improved status. As many anthropologists have noted, including Douglas (1966), Smith (2000) and Modan (2007), dirt is a signifier of imperfection and inferiority, leading to the criminalisation of people and spaces as impure and contagious. Notions of impurity inform a moral geography that associates space with embodied markers of difference, including race or style of dress, or even presence in certain public places. Bodies that occupy impure spaces are interpreted as immoral, and vice versa.

La Ciénaga's moral geography classifies the private sphere as more moral than the public sphere.[6] While La Clarín has an active street life, abajo is infamous for its public activities in sites such as *colmadones* (open air bars), whose noise and revelry mark them as belonging to the 'popular' class. The rowdiest colmadones are located off the main streets and draw their clientele from neighbours who live close by. Colmadones occupy a similar place in Dominican social imaginaries as dancehalls do in urban Kingston (see Stanley-Niaah 2004, Thomas 2004). They are places for the (relatively) young and (somewhat) disreputable, particularly men, to gather, drink beer and rum, listen to reggaetón and bachata and flirt. Every weekend, the sounds of partying float up to La Clarín from abajo, occasionally accompanied by gunshots. Often the activity is still audible the following morning. 'Listen to the *tígueres*! (tigers)' residents regularly exclaimed, 'Listen to the state of *desorden* (disorder) that this country is in!' For residents arriba, the partying abajo indicates the deterioration of the Dominican people, who have fallen into *corrupción* (corruption or immorality) in hard times, rather than persevering with traditional peasant values of hard work and austerity.

The moral geography often relates space of residence to region of origin or racial characteristics, which are interpreted as cultural and moral differences. Approximately 60 per cent of residents originated from the arid southwest of the country, 22 per cent from the fertile agricultural centre of Cibao and just 5 per cent from the southeast, the country's largest tourist region (Tejeda 2000: 33). Residents from Cibao were the first to arrive in the 1960s and 1970s, occupying the barrio's high ground, arriba. They tend to be light-skinned because Cibao was primarily settled by European migrants. Darker-skinned residents from the southwest arrived in the 1980s and built their houses in the swamp abajo, reinforcing association of dark skin with poverty.

There are few Haitians in La Ciénaga, yet to some residents they appear numerous, particularly in times of economic insecurity when illegal migration most appears to be a threat to Dominican livelihoods. Fear of cultural difference, violence and irrational ideas that Haitians are primitive are parts of the racist discourse. State racism classifies both blackness and whiteness as foreign, but while blackness signifies cultural difference, poverty and criminality, whiteness signifies privilege and wealth. In La Ciénaga this state racism is reflected in notions of what Haitians are and what they do, including perceptions that they are poor, illegal immigrants; that they therefore mostly live abajo; and that they

6 Depending upon other factors; these classifications are not absolute.

are uncivilised and dangerous, liable to violent outbursts. This is by no means the only view of Haitians, nor could it even be argued that most people believe these stereotypes; rather, they are convenient tropes that are drawn upon in times of trouble. In daily life, relations between Dominicans and Haitians are characterised by a good deal of cooperation.

Economic crises can increase the temptation to scapegoat Haitians, but it can also break down distinctions if the poor cannot afford to distinguish themselves from each other. One Haitian friend of mine commented how Dominicans have begun to eat certain affordable foods that they used to disdain:

> Before, Haitians had no value, if you lived here they treated you like a rag because you weren't in your own country. They used to say that *arrenque* (maize flour), black beans, cornflower and sardines were Haitian foods. They didn't eat these things. Bananas were Haitian food. Now there are vitamins in them! Arrenque is vitamin and nourishing, all this nourishes. Before if you ate this you must be Haitian, but now, ooh! Now it's tasty for them because things are bad everywhere. (Amélie)

Distinction is therefore broken down by economic necessity as people find they must engage with material conditions differently in order to cope with crisis. Amélie, a Haitian woman who migrated to La Ciénaga in 1981, has done the reverse: her engagements with the material environment have been orientated towards improving her status, although they have also highlighted her difference. Amélie most strongly felt her status as an outsider when she and her husband constructed their new house. Originally, Amélie and her sister settled in different houses located abajo, but around five years ago Amélie and her husband secured a plot of land on the edge of La Clarín. It is a tiny three-room concrete house that they built themselves with money her husband earned selling clothes in the city market. Their neighbours were jealous that Haitians could own a 'good' house, while many Dominicans could not. Recounting an incident on a sugar plantation where a Haitian *bracero* (sugarcane cutter) had entrusted his savings to a Dominican who subsequently betrayed him, Amélie defended her right to belong:

> If a Dominican sees that a Haitian has a good house, you know what they do? They look for trouble with them, they kill them or whatever, so that they can take it from them … If I struggle and manage to build my house then nobody should abuse me or look for my [immigration] papers. Leonel's [Fernandez] government should take account of this. (Amélie)

Here Amélie draws a clear link between the treatment of Haitians at the national level and their ability to belong in local communities. She suggests that if the Dominican government did not harass Haitians, they could command the same respect as Dominicans for having struggled to achieve a good house. Instead, Amélie feels that her family's position in the barrio is threatened due to state discrimination. Their house is located on low ground but it can only be accessed via the high ground of La Clarín, so Amélie legitimately claims to reside arriba. She takes pride in her La Clarín residence and the status it offers her, particularly since her 'good' house and her residential location allow her to defy common stereotypes of Haitians as occupying the worst parts of the barrio and the worst houses.

Amélie's position is similar to that of the Dominicans around her in that she is a migrant who must invest in the production of locality when her aspirations and sense of belonging lie elsewhere. But unlike Dominicans in La Ciénaga, who can move into mainstream society if they have the resources to do so, Amélie would always be recognised as an outsider. Given this restriction on lateral movement, and her unwillingness to return to Haiti, Amélie has a greater interest in the production of locality as a strategy to realise belonging and socioeconomic mobility. Her construction of a house is a symbol of her commitment to 'local' values and indeed their universality for 'respectable' people everywhere, in resistance to popular discourses that cast Haitians as primitive and backward. Her extensive community work with the Catholic church involves her movement across all parts of the barrio, breaking down the moral geography in which certain places and people are labelled deviant.

This is not just Amélie's strategy, but also that of the church, which has an interest in consolidating the Catholic community, particularly in the face of a rising Pentecostal presence. The Catholic church attempts to foster a sense of unity by calling on a global sense of shared humanity in which all people are equal under God. It has active programs to counter racism and other forms of discrimination by pointing to a higher form of authority than the Dominican state, which has been historically involved in the production of cultural difference. Along with the rootedness of sociality in the domestic sphere, this global morality is presented as an antidote to the deterioration of social structure originating from the transnational sphere in the form of economic crises and the individualism that mass consumption brings. In the church's view, social consolidation within localities is imperative to counter the corrupting effects of globalisation and to challenge the inequalities that produce residents' reluctant commitment to living in the community. In their view, the barrio can in fact be transformed into a modern urban community whose residents identify socially and culturally with the locality that they produce, rather than merely a choice of the necessary. But whether the politics of place concerns staying or going, its reference points will continue to be located outside the community as much as its strategies are enacted on the inside.

Conclusion

Residents of La Ciénaga face what appears to be an impossible problem. They have been ascribed by circumstance to a small geographical locality, which then forms the basis of their daily struggles for survival and their negotiations for a better social position. Over the past few decades they have made remarkable strides in forging a community and strategies which use their localisation to create a permanent position in the city. In doing so they have acted counter to the wishes of the state and the general public who cast them as illegitimate. However, since the barrio was founded a few decades ago, residents have primarily utilised locality politically to gain inclusion into the city and a progreso that is both national and transnational, through the construction of the built environment and their ability to spatialise social relations within it.

The politics of place may rest upon the production of locality, but its reference points are the state and the transnational sphere. This is indicative of how residents' lives and livelihoods depend upon forces outside the locality, despite the fact that it is these very forces that have localised them so thoroughly. The politics of place reflect the wide range of interests within this complex field, how parameters for the types of unity that develop are set and how much commitment to progress within place can be procured. Understanding people's engagement in place politics – individual and collective – requires examination of their relationships and positioning beyond the locality upon which that action is centred. This is not to argue that locality has no agency of its own, but rather that people are never merely local: 'Everyone can relate themselves (or is allocated) to a multiplicity of spaces – phenomenal and conceptual – whose extensions are variously defined, and whose limits are variously imposed, transgressed, and reset' (Asad 1993: 8). The production of locality, for cienigüeros, has always been, and will continue to be, a strategy for mobility *out* of the barrio as much as it has been a way of building a life project *within* it.

References

Asad, T. 1993. Introduction, in *Genealogies of Religion: Discipline and Reasons of Power in Christianity and Islam*, edited by T. Asad. Baltimore: The John Hopkins University Press, 1–24.

Bauman, Z. 1998. On glocalization: or globalization for some, localization for some others. *Thesis Eleven*, 54, 37–49.

Bosch, J. 1992. *Clases Sociales en la República Dominicana*. Santo Domingo: Editora Corripio.

Bourdieu, P. 1984. *Distinction: A Social Critique of the Judgement of Taste*. Cambridge: Harvard University Press.

Derby, L. 2009. *The Dictator's Seduction: Politics and the Popular Imagination in the Era of Trujillo*. Durham and London: Duke University Press.

Douglas, M. 1966. *Purity and Danger: An Analysis of Concepts of Pollution and Taboo*. London: Routledge.

Escobar, A. 2001. Culture sits in places: reflections on globalism and subaltern strategies of localization. *Political Geography*, 20, 139–74.

Fay, M. and Wellenstein, A. 2005. Keeping a roof over one's head: improving access to safe and decent shelter, in *The Urban Poor in Latin America*, edited by M. Fay. Washington, DC: The World Bank, 91–124.

Ferguson, J. 2003. Migration in the Caribbean: Haiti, the Dominican Republic and beyond. Minority Rights Group International.

Grasmuck, S. and Pessar, P. 1991. *Between Two Islands: Dominican International Migration*. Berkeley: University of California Press.

Gregory, S. 2007. *The Devil Behind the Mirror: Globalization and Politics in the Dominican Republic*. Berkeley: University of California Press.

Harvey, D. 2000. Cosmopolitanism and the banality of geographical evils. *Public Culture*, 12, 529–64.

Hoffnung-Garskof, J. 2008. *A Tale of Two Cities: Santo Domingo and New York After 1950*. Princeton and Oxford: Princeton University Press.

Holston, J. 1991. Autoconstruction in working-class Brazil. *Cultural Anthropology*, 6, 447–65.

Martínez-Vergne, T. 2005. *Nation and Citizen in the Dominican Republic, 1880–1916*. Chapel Hill: The University of North Carolina Press.

Miller, D. 1994. *Modernity: An Ethnographic Approach: Dualism and Mass Consumption in Trinidad*. Providence and Oxford: Berg Publishers.

Modan, G. 2007. *Turf Wars: Discourse, Diversity, and the Politics of Place*. Malden, Mass.: Blackwell.

Robotham, D. 1998. Transnationalism in the Caribbean: formal and informal. *American Ethnologist*, 25, 307–21.

Smith, D. 2000. *Moral Geographies: Ethics in a World of Difference*. Edinburgh: Edinburgh University Press.

Stanley-Niaah, S. 2004. Kingstons's dancehall: a story of space and celebration. *Space and Culture*, 7, 102–18.

Taylor, C. 1994. *Multiculturalism: Examining the Politics of Recognition*. Princeton, N.J.: Princeton University Press.

Taylor, E.B. 2009a. From *el campo* to *el barrio*: memory and social imaginaries in Santo Domingo. *Identities: Global Studies in Culture and Power*, 16, 57–78.

Taylor, E.B. 2009b. Poverty as danger: fear of crime in Santo Domingo. *International Journal of Cultural Studies*, 12, 35–52.

Taylor, E.B. 2009c. Modern *dominicanidad*: nation-building and politics of exclusion in Santo Domingo since the 1880s. *Dialectical Anthropology*, 33(2), 209–17.

Thomas, D. 2004. *Modern Blackness: Nationalism, Globalization, and the Politics of Culture in Jamaica*. Durham and London: Duke University Press.

Vicioso, S. 2000. Dominicanyorkness: a metropolitan discovery of the triangle. *Callaloo*, 23, 1013–16.

Wilson, P. 1973. *Crab Antics: The Social Anthropology of English-Speaking Negro Societies of the Caribbean*. New Haven and London: Yale University Press.

Wolf, E. 1967. Caudillo politics: a structural analysis. *Comparative Studies in Society and History*, 9, 168–79.

Chapter 7

Little Anglo-India: Making Australia 'Local' at St Joseph's Hostel

Robyn Andrews

On first entering St Joseph's Hostel[1] on a late winter morning in 2004, I immediately noted the set of pictures depicting the Stations of the Cross lining the hallway, the artificial banana tree – complete with a toy monkey hanging from the fronds – in one corner of the main living area, the unmistakable scent of curry and the sound of western 1950s songs playing on the radio. These, along with the presentation of a cup of sweet milky tea and an invitation to stay for a potluck[2] curry lunch and a round of afternoon Housie,[3] were hallmarks of Anglo-Indian living and hospitality that I had become familiar with during my PhD research, an ethnographic exploration of Kolkata's Anglo-Indian community (Andrews 2005).[4] However, instead of rickshaws and incense, the streets outside the doors of St Joseph's were lined with Eucalyptus trees; this residential care facility for older Anglo-Indians is located in suburban Melbourne, making it the only Anglo-Indian rest home outside of India.

Anglo-Indians are of mixed European and Indian descent, originating in India primarily as a result of a British presence, beginning with trade relations in the 1600s with the establishment of the East Indian Company, and then rule from 1858. Due to filial links with the British (and other Europeans) and also, arguably, their exclusion from caste-conscious Hindu society, they adopted a distinctly Western way of life. Following India's independence from Britain in 1947, the Anglo-Indian community underwent rapid migration, primarily to Britain and commonwealth countries including Canada, New Zealand and Australia. In the course of my research I discovered St Joseph's, a little Anglo-India almost on my doorstep, and I was fascinated by how the rest home achieves its 'Anglo-Indianness' through a combination of décor, food, rituals, relationships and social engagement, all of which resonate with pre-migration experiences of being Anglo-

1 This institution is most often referred to as St Joseph's Hostel, although the name on the front of the building is 'St Joseph's Centre. Older Persons Hostel'.

2 An Anglo-Indian 'potluck' is usually an impromptu invitation to dine on whatever happens to be cooking in 'the pot', rather than a request to contribute to the meal.

3 'Housie' is an Anglo-Indian name for 'bingo'. Both terms are used interchangeably.

4 Kolkata was previously known as Calcutta.

Indian in India. The hostel also serves as a centre for Melbourne's wider Anglo-Indian community. This community of approximately 4,500[5] people continues to enact practices and activities common among Anglo-Indians living in India, including the maintenance of a tight-knit family life, religiosity, holding dances on occasions including The May Queen and New Year's eve, cooking particular cuisine such as kul kuls and 'ball curry', and team sporting activities, particularly hockey and Pagal Gymkhanas.

This chapter is written from research I conducted at St Joseph's over two intensive phases of fieldwork[6] in 2006 and 2007,[7] and focuses on the day-to-day practices of its residents, the hostel's relationship with the wider community and the ways in which the facility has been made into a distinctly Anglo-Indian home, producing a 'local' space in the adopted environment of Melbourne. I use the work of Arjun Appadurai as a framework, and examine how those involved in St Joseph's have managed to 'produce locality' both as a physical space and as a 'structure of feeling' (Appadurai 1996: 178–9). Appadurai's focus on the production of locality privileges relational and contextual dimensions over ideas relating to the materiality of physical spaces, which he refers to as 'neighbourhoods' but which include both spatial and virtual varieties (Appadurai 1996: 179). An application of his theory to this Anglo-Indian case study provides an example of how materiality can be central to the production of locality.

Appadurai's attention on the globalised context – that of the deterritorialised, diasporic and transnational (1996: 188) – was developed as a result of, and as a way of making sense of, the new world order where people are increasingly likely to experience physical mobility (whether voluntarily or forced) within their lifetimes, where diasporic communities are becoming commonplace and where the mass media reaches out to people and provides them with 'a wider set of possible lives' to consider, 'some of which enter the lived imaginations of ordinary people' (1996: 53) and affect their imagination of alternative social worlds, if not their life trajectory. The perspicacity of a theory focusing on a sense of locality rather than on physical or material dimensions of neighbourhoods is that it can account for more people's experiences within a range of possible social situations, or alternative social worlds, that occur for people in today's globalised world.

---

5 Based on census figures for the State of Victoria, of which Melbourne is the state capital, which listed 4,622 people of Anglo-Indian descent in 2001, with 12,327 Anglo-Indians in the whole of Australia. As Anglo-Indians have continued to migrate from India to Australia, these figures will be higher now.

6 My research methodology included intensive participant observation (including one 36-hour stay); formal interviews with residents, often collecting life histories where longer interviews were possible and informal conversations with staff, management, residents and visiting family members.

7 This is part of a larger project on Anglo-Indian ageing. I have also carried out (in 2007 and 2009) similar participant observation research in Kolkata's three rest homes which are dedicated to the housing and care of older Anglo-Indians.

For Appadurai, 'neighbourhood' signifies 'the actually existing social forms in which locality, as a dimension or value, is variably realised' (1996: 179). Neighbourhoods may or may not be physical, but are characterised by 'actuality' and 'potential for social reproduction' (Appadurai 1996: 179). They are the places where a sense of locality can be produced, and reproduced, but which differ in fundamental ways from locality. 'Locality', for Appadurai, is a 'structure of feeling' (1996: 182) and as such is always further abstracted from 'actuality' than neighbourhoods are. He posits that social scientists should not study the production of neighbourhoods as ends in themselves (1996: 180–82); rather, they should see neighbourhoods as providing the opportunity or context for the production of locality, which, he stresses, is a fragile and difficult achievement (1996: 198). Appadurai's way of viewing the relationship between neighbourhood and locality is illuminating in the context of this Anglo-Indian hostel. He would posit that while the materiality of the hostel, along with particular rituals and ways of socialising are important, it is what they generate (locality), that is the really significant expression. The case study presented here, however, indicates that his distinction between neighbourhood and locality is much harder to tease out than he proposes.

Appadurai also writes of locality as having a 'taken-for-granted, commonsensical, habitus dimension' (1996: 185) and argues that while 'the transformation of spaces into places requires a conscious moment', it 'may subsequently be remembered as relatively routine' (1996: 183). This insight sheds light on the 'off-hand' and unreflexive explanations I was offered in relation to the construction of St Joseph's, especially by the hostel's founder: he was unable to articulate a specific ambition of reproducing in this hostel the Anglo-Indian rest homes he had known in India.

Locality does not require a physical neighbourhood, it can, for example, exist as a virtual site (1996: 174, 195). Where a neighbourhood exists, however, it is the practices performed within it that produce what Appadurai sees as being of fundamental importance. Neighbourhoods, being 'life worlds constituted by relatively stable associations, by relatively known and shared histories, and by collectively traversed and legible spaces and places' (1996: 191) produce the structure of feeling that connects people to each other and engenders a sense of being part of each others' community, and (particularly where the neighbourhood is spatial rather than virtual) of being together in a place of 'their own'. The construction of a distinct space and way of life within the Anglo-Indian home that I discuss in this chapter draws on both material and non-material dimensions in order to produce the distinctly Anglo-Indian 'locale' that it is. In this chapter I also examine my findings in the context of related literature on retirement communities and the elderly, as well as drawing some comparisons with Anglo-Indian experiences of living in rest homes in India.

Anglo-Indians in India and Abroad

Anglo-Indians are defined in the Indian Constitution, Section 366 (2) as follows:

> a person whose father or any of whose other male progenitors in the male line is or was of European descent but who is domiciled within the territory of India and is or was born within such territory of parents habitually resident therein and not established there for temporary purposes only.

Despite living in India, most Anglo-Indians identified largely with British cultural norms, speaking English as their mother tongue, practicing Christianity and adopting western styles of dress and cuisine. This may, in part, be due to their exclusion from the social order created on Hindu ideology, which is based on a particular sense of ritual purity and pollution and does not tolerate physical intimacy between members of different castes, let alone different religions. As a result, any man or woman involved in such an exogamous relationship, and progeny from such a relationship, are viewed as polluted and are placed outside the caste system. In India under colonial rule, Anglo-Indian employment tended to be mainly within the infrastructures that maintained British India – post and telegraph, customs, police, teaching, nursing and the armed forces. A large number also worked for the railways, to the extent that Laura Bear wrote about them as 'the railway caste' (Bear 1998: 158–91).

Once India gained independence in 1947, Anglo-Indian emigration began to rise for a range of reasons. Initially, there was fear of reprisals from the newly independent Indian government because of their close links with, and loyalty to, the British Raj. It soon became evident that these fears were unfounded, but they were replaced by anxieties caused by the threat of Hindi replacing English as the national language and the departure of many foreign-owned and run businesses that had employed Anglo-Indians. More recently, particularly when the conservative Bharatiya Janata Party (BJP) was in power from 1998 to 2004, Hinduisation processes have left many Anglo-Indians feeling they may not have a future within India. In my time spent in Kolkata over the last eight years, I found many Anglo-Indians, who have always been a minority, losing their sense of being 'at home' there. Some reported experiencing increased hostility solely on the basis of their heritage, which may account for the continuing stream of Anglo-Indian migration out of India. Currently, the number of Anglo-Indians residing out of India is believed to exceed those remaining in the subcontinent, which is estimated between 200,000 and 300,000. However, as a number of researchers and Anglo-Indian leaders have noted, the exact number of Anglo-Indians in India is difficult to determine since they are no longer enumerated separately in the 10-yearly national census (see Anthony 1969, Younger 1987, Caplan 1995, Bear 1998, Williams 2002, Blunt 2005).

As a commonwealth nation, Australia has been a popular destination for Anglo-Indian migrants. However, shifting attitudes and political policy towards race in

Australia has meant a diverse history in terms of its treatment of Anglo-Indian migrants (see Blunt 2005, Lewin 2005). For the first half of the twentieth century, Australia adopted a racially restrictive and effectively 'whites only' immigration policy. This policy began to change with '[t]he revised Migration Act 1958 [which] stated that people who *appeared* to be less than 75 per cent European should be rejected as migrants (Jordens 1995). This allowed the migration of people of "mixed race" such as the Burgers of Sri Lanka, Anglo-Indians, Anglo-Burmese ...' (Lewin 2005: 635, emphasis added). However, Anglo-Indians are known to have highly variable physical characteristics, with anecdotal accounts of both swarthy and fair siblings within the same family. This policy effectively meant that Anglo-Indians who were fair enough to 'pass' were admitted, while others – possibly even siblings or people with an actual higher percentage of European ancestry – with darker skin were excluded. From 1964 onwards Australian immigration policy became more inclusive, but still favoured Anglo-Indians over other Indian groups as it was assumed that because they were already proficient in English and shared 'western' cultural norms, such as Christianity, they would assimilate easily, making them 'good migrants' (Blunt 2005, D'Cruz 2006). In 1990, the Bob Hawke-led Labour government shifted from assimilationist migration policies to encourage diversity and focus on building a more multicultural Australia.

St Joseph's Hostel[8]

The founder and current manager of St Joseph's Hostel recognised the opportunities presented by the multicultural rhetoric and policy that began to take hold in the 1990s; through the Victorian Anglo-Indian Association (established in 1971 and now renamed the Anglo-Indian Australasian Association of Victoria), a proposal was put forward for a residential care facility, to be partly funded by the Australian government. The application was successful, and in 1994 St Joseph's Hostel opened in a Melbourne suburb that, at the time, had a significant Anglo-Indian population.

The hostel had 16 bedrooms, each equipped with an ensuite bathroom. It was a low care facility with no resident medical staff and was therefore suitable only for residents who were relatively healthy and mobile and did not suffer significantly from age-related dementia. Since then, in 1998, St Mary's retirement village was constructed on an adjacent piece of land and offers one and two bedroom self-contained units. This is also an initiative of the Anglo-Indian Australasian Association of Victoria, which offers the units on a rental basis and manages the properties. The units are available only to Anglo-Indians and have been consistently fully occupied since they were completed. There are plans to build more units for similar independent living on land the association owns on the other side of St

8 In Australia the term 'hostel' has the same meaning, in this context, as 'rest home' does in New Zealand and other countries, referring specifically to a low care, non-hospital facility for senior citizens.

Joseph's Hostel. Both the hostel and the retirement units are just a short walk from a medical centre, a suburban shopping centre and the railway station.

The single-storied, orange-brown brick and mortar exteriors of St Joseph's and St Mary's are typical of suburban Australia, and when empty the hostel rooms are spacious, with pale painted walls, single beds and ensuite 'wet-area' bathrooms. The buzzer call-buttons in the bathroom and above the bed and the push-in timer button for the bathroom lights create a utilitarian, institutionalised feel to the rooms. However, the rooms that were occupied were transformed by personal touches, many of which greatly contributed to the distinctly Anglo-Indian atmosphere: altars set up on dressing tables, crocheted doilies, jars of Horlicks and photographs – many taken in India. The living area also featured little Anglo-Indian touches; the aforementioned plastic banana tree is reminiscent of the real ones I saw in one of Kolkata's rest homes.

The hostel's founder told me that while living in, and later visiting India he had been to Anglo-Indian rest homes in Chennai.[9] However, when asked to elaborate the ways that St Joseph's provided a specifically Anglo-Indian ambience, he only emphasised the food they serve. This implies that while many of the details I have noted had not been consciously orchestrated, those Anglo-Indian spaces and day-to-day practices had been replicated here as the most suitable and 'natural' environment for older Anglo-Indian migrants. The decisions on how the home was to be set up and run were therefore doxic (Bourdieu and Wacquant 1992: 73), with the environment produced by the knowledge and experiences of the founder and the others involved in the home creating a space that felt familiar despite the foreign setting. Appadurai writes about the changing forces acting on habitus in the globalised world, focusing on acts of 'improvisation' which were traditionally theorised as acting against habitus: 'where once improvisation was snatched out of the glacial undertow of habitus, habitus now has to be painstakingly reinforced in the face of lifeworlds that are frequently in flux' (1996: 56). Even if not consciously orchestrated, it was important to the founders to achieve what Erin Taylor refers to as the 'production of locality', which she describes as 'the development of a relatively bounded material environment and the objectification of the social relations in it' (Taylor this volume). In the rest of this chapter I tease out the specifics of the material and ritualised ways in which this has been achieved, and demonstrate how engrained the relational and contextual spheres are in the materiality of this urban space.

9 There are dedicated Anglo-Indian rest homes in all of the main centres of Anglo-Indian population in India, including Kolkata, Lucknow, Bangalore, Delhi and Chennai. There are also a number of church-based rest homes throughout India that older Anglo-Indians may choose to reside in.

Home Food

The food served at St Joseph's was clearly identified by residents, staff and visitors as something that set this facility apart from other Australian retirement homes. Denzil,[10] a 92-year-old who had left life as a railway station manager in India over 30 years ago to bring his family to Australia, even identified food as the primary reason he had moved into St Joseph's. 'The food. That's what I came here for … not for anything else, but I only came for the sake of the food. I told you that before, because I couldn't eat Australian food.' For Denzil, who permitted me to conduct longer life history interviews, food had always been key and had even played a large part in his experience of migrating permanently with his wife and younger children to join their adult daughter in Australia. Before leaving India, they bought '… such a lot of stuff: we brought garlic, ginger, we brought chillies, you know, dried chillies. I went to the mills and got it ground. We filled up tins and came with it all'. However, Denzil was far from alone in stressing the importance of food. Mary, who is 79 and left India to come and help her daughter look after her young grandchildren, had been in the hostel for five of her 15 years in Australia. She said, 'The food is a big thing. And that's a, yeah, a great thing. Most of the residents are happy because they say it's like eating home food'.

Mary's mention of 'home food' is significant; while in Kolkata, I was always told that 'home food' was best, and safest, so it was always preferred to food from 'outside'. (These opposing tropes are further complicated in the Indian setting where the 'home food' is invariably cooked by people from 'outside', that is, Muslim or low caste Hindu servants who must be trained to cook dishes which are unfamiliar to them.) I was also told that Anglo-Indian food preferences were largely western ('An Anglo-Indian will always prefer a Continental dinner.' (Chattopadhyay 2007)), but what Anglo-Indians said they loved most of all about this Melbourne home was the specifically Indian flavours in the food – the curries, rice and *dâl* that they are served every day. Much like their heritage, it seemed evident that the elderly Anglo-Indian residents of St Joseph's preferred a mixture of western and Indian in their cuisine. For instance, breakfast in the hostel is toast, cereal or porridge with tea and coffee (rather than an Indian breakfast such as idli or dosa); lunch is a tasty and generous meal that always includes several different curries – usually at least one type of chicken curry, and a vegetable one, rice, *dâl* and salad, followed by fresh fruit. Dinner is soup (which was quite spicy on the night that I stayed over) or a pasta dish, or chicken pieces with cooked vegetables, followed by jelly or a custard-based dessert.

One particular fieldwork vignette reveals the role that food plays in St Joseph's for linking these migrants to their history in India. At the end of a two-week research stint, where I had spent the majority of each day in the hostel, I asked the cooks for advice on how to best give residents a small token of my thanks for their participation and patience. As many of the residents were diabetic, even though most seemed to

10 All names are pseudonyms.

have a weakness for sweet things, the obvious – biscuits or chocolate – were not an option. Instead, I was advised to buy a tray of ripe mangoes. The following excerpt from my field notes picks up from my last lunch at the hostel:

> By the time we come into the dining room some are already sitting at the tables with their lunch. Today it's sausage curry, *kofta* curry with potatoes, *dâl* and rice, a salad of grated carrot and red onion, turmeric cabbage, and a *brinjal* pickle. There are glasses of water on the tables, along with serviettes, hanging sets of cutlery, plastic flowers, and packs of cards. Dessert is always fresh fruit: today it's mangoes – my contribution. Earlier I'd given the tray of mangoes to Laura, the Anglo-Indian cook, but she said I would have to distribute them, and find out how the residents wanted to eat them. I was puzzled – how many ways are there of eating a mango? So I asked each in turn. David asked for Laura to cut his – and wanted only the 'cheeks'. Gerry wanted to do his own, so I got a plate and knife for him. Mary wanted her cheeks diced. Ken's snuff-box cut didn't work. He ended up with something more like mashed mango. Some took theirs back to their rooms – 'for later'. I heard a story of a trip David had made as a school boy, travelling with his father in the cabin of a freight truck to Salem – 'where the best mangos come from'. Denzil repeated an oft-told refrain, that nothing tastes as good as it did in India. Astonishingly, most could tell me when they'd eaten their last mango. (Andrews 2008)

This experience is consistent with other migrant stories of resettling – it is often not until good supplies of 'familiar' food are assured that migrants really start to feel settled in their adopted country. Migrant anecdotes include, for example, Italians who were dismayed at their inability to obtain good coffee in the antipodes, Indians who couldn't get the spices they were used to (Patel 1990: 10) and Germans hungering for good bread in New Zealand (Bönisch-Brednich 2002: 33). The link between taste, smell and memory is well established (see, for example, Sutton 2001), and there are anthropological accounts of the significance of food to memory and to re-establishing a sense of being 'at home'. Paul Stoller writes of the importance of the senses, and advocates the need to include taste in ethnography (1989). A similar phenomenological approach is encouraged by Michael Jackson (1996), who calls for a sensitivity to the sensory experiences of research participants. This may be particularly important for rest home residents who no longer have a home that is specifically their own. Bland, who wrote a critical ethnography of New Zealand rest homes, noted that 'residents generally preferred, and felt more comfortable with the food they had grown up with', adding, 'A New Zealand retirement home dismissed its chef after he allegedly committed what were described as "several food faux pas" (Grunwell 2000: 6), including failing to boil vegetables "to destruction", refusing to make plain food, and using too many spices' (Bland 2004: 118–19). In the Anglo-Indian context of St Joseph's, of course, the faux pas would have been the opposite – too few spices would be sure to elicit complaint.

Along with the food, it is important to consider the significance that mealtimes hold for rest home residents. Shared dining provides an opportunity to socialise, and fill up or mark passing time during a quiet day. In St Joseph's, some differences between a distinctly Anglo-Indian institution and other facilities, such as the New Zealand ones Bland conducted research in, are played out through food and hospitality. Contrary to Bland's findings that residents, particularly women, were frustrated at not being able to fulfil the role of hosts (Bland 2004: 118), in St Joseph's residents were able to help themselves and their visitors to tea or hot drinks at any time they wished, contributing to a sense of making the hostel their own. Notably, some residents also kept a supply of Horlicks (a malted drink, originally from the United Kingdom but commonly produced and sold in India since the 1950s). As part of their evening routine, they would fill a thermos with hot water and make a hot Horlicks drink in their rooms later; this is a practice they are likely to have been familiar with from living in India.

Another feature of mealtimes that sets St Joseph's apart from other, more 'institutionalised' Australian facilities for the elderly is the fact that many residents play a part in preparing meals. Several of the female residents, for example, were involved in assisting the two employed cooks to prepare fruit and vegetables for the meals. One man made it his duty to set the table and pour water into glasses in preparation for each mealtime. Again, this contrasts with Bland's findings; she notes that residents keenly felt the 'loss of their social roles in association with food provision' (Bland 2004: 118). This involvement with the day-to-day responsibilities of keeping the hostel running was not restricted to meals. One man, for example, took it upon himself to open one side of the dining room curtains in the morning, while another man took care of the other side. Another, who was last out of the lounge-cum-dining room at night, would run a red pen through the day's date on the calendar as he left to retire for the night.

While the ability to perform some functions as a resident is unlikely to have been a reason for coming to the hostel in the first place, it was important in allowing the residents to feel comfortable and 'at home' once they were there. The familiar food and ability to treat the hostel as their home contributes significantly to the development of a particularly Anglo-Indian 'structure of feeling' creating a sense of this being 'their place' within the materiality of this specific space. The management deserves credit for enabling residents to perform these valuable roles, but it is the residents' initiative, which has been encouraged, rather than the management having actively organised residents to take responsibility for these tasks. Therefore the residents must be seen as agents, actively contributing to the culture of the hostel, which builds a sense of locality. This active and independent nature of St Joseph's residents is further explored next.

Christian Materiality

I have already noted the obvious presence of Christian iconography and religious symbols throughout St Joseph's. In addition to the Stations of the Cross along the corridor, the dining room walls were hung with holy pictures donated by Anglo-Indians who had been involved in the early days of setting up the hostel. In the privacy of residents' rooms, this theme continued: often, holy pictures were among photographs of grandchildren, family weddings and graduation ceremonies, and elsewhere there were Virgin Mary-shaped bottles of holy water from pilgrimage sites in India, or rosary beads beside beds and crucifixes hanging on the walls.

Not all Anglo-Indians are Roman Catholic, but even among Protestant Anglo-Indians there are practices – such as prominently displaying religious icons – that are generally characteristic of Catholic practice. The distinct materiality of Anglo-Indian Christianity in India struck me early on in my research in Kolkata. I had come to recognise an Anglo-Indian home by two items of furniture virtually always present: an altar (a small platform mounted on the wall at about shoulder level with candles, statues and other religious objects arranged on it) and a showcase (a cabinet with glass doors, used to display precious objects such as souvenirs, small framed photographs, crockery and decorations). The displayed objects in the public and private spaces of the hostel mirrored these observations I had made in India.

June, a Catholic resident of eight years who came to Melbourne to live with her daughter after she was widowed, commented, 'It's good to have the Stations of the Cross here, especially coming up to Easter'. In India, Christianity permeates all areas of Anglo-Indian life: conversation is sprinkled with references to their religiosity, homes are adorned with pictures and icons, and every week includes visits to churches and plans for pilgrimages. For many, their primary and secondary education was conducted in church-based schools. One woman I spoke to drew comparisons with Anglo-Indian practice of religion out of India saying:

> We, Indians, are all spiritual people. We practice different … worship different Gods, but we all do so fervently. People say we do so because we've got nothing. That we pray to have a better life, and if we had the sorts of lives they have in the west then we wouldn't need our religion. They say that, they do! Someone said it to me.

She continued by stating sadly that her brother and sister who have left India 'left for a better life, but I don't know that that's what they've got. Materially it's better, yes, but spiritually …' She knows (or strongly suspects) that none of her overseas family practise their faith as ardently as she and her brother and most Anglo-Indians in Kolkata do. While there may be a lessening of the overt practice of faith in Anglo-Indian homes in the west, in this hostel it was very much in evidence, and was comparable to the Kolkata rest homes I spent time in.

St Joseph's also incorporated Christian ritual into the residents' everyday lives. Prayer groups met each Tuesday morning with an Anglo-Indian minister who would also come to give communion to Anglican residents, and local Catholic Anglo-Indians gathered at the hostel once a week to lead the Catholic residents in the Rosary. Occasionally mass was celebrated in the hostel and there was usually a mini bus available to transport residents who wished to attend Sunday services.

Recreation and Sociality

As is the case in most rest homes – Anglo-Indian or not – residents have recreational activities arranged for them. In St Joseph's there are two afternoons of bingo a week (with announcements such as 'jeldi paunch' – Hindi for 'quick five' – for the first to mark five numbers), and on another afternoon an Anglo-Indian group come for a sing-along of 'old favourites'. Both of these activities are typical of Anglo-Indian entertainment and include elements such as the use of Hindi. In Kolkata I attended 'house parties', many of which included a sing-along, and most larger scale community-run daytime social events, such as Christmas parties, would run a bingo session. Despite not identifying these as specifically Anglo-Indian practices, by scheduling these activities the manager provided an environment similar to what residents may have experienced in India.

Almost all of the residents' social interactions are with other Anglo-Indians: from the cooks and other paid staff, to the volunteer nurse, to the Anglo-Indian ministers and priests, to the visitors that the hostel attracts – relatives and friends, the sing-along group and outsiders who come to play bingo. For these elderly residents, some of whom are beginning to lose their hearing and whose diction is not always clear, there must be significant advantages in mixing with people who have similar accents, as well as cultural expectations.

Residents' shared experiences and similar histories are also a major factor in facilitating easy communication and establishing friendships. All of the Anglo-Indian residents had lived in India and made the decision to migrate to Australia as adults. Over the course of my fieldwork visits I was party to a number of discussions which illustrated their shared understanding: I heard discussions about recipes and food, family Christmas traditions, comparisons between India's summer heat and Melbourne's and discussions of the upcoming World Anglo-Indian Reunion. Although English is their mother tongue, Hindi words were regularly dropped into conversations, as were colloquialisms that I recognised from Kolkata. For instance, they talked about 'settling their room' (meaning to tidy the room), attending a 'Christmas Tree' (referring to a Christmas party) or about someone becoming 'friendly' with someone else (in India, this term signifies romantic involvement). These types of knowledge systems and rituals of exhibiting knowledge are crucial for demonstrating membership in the local group. A comparable parallel is the Jewish Senior Citizens' Centre in California, which Barbara Myerhoff describes in her 1978 ethnography, *Number Our Days*. The Centre functions as the daytime home to a group of 'former migrants from Eastern Europe now mostly abandoned by their more or less assimilated New World progeny' (Turner in Myerhoff 1978: xiv). It offers a range of services from acting as a halfway house, to cooked meals and cultural activities, but its over-arching function is to provide an opportunity for immersion in a Yiddish environment. Myerhoff notes that its significance is in residents being able to 'indulge their passion for things of the past, [and] enjoy Yiddishkeit without fear of being stigmatised as "not American"' (Myerhoff 1978: 9).

For Anglo-Indians, stigmatisation was not stated as being a reason for establishing the hostel, although in the life stories I collected this was sometimes included as a factor in their settlement experiences in Australia. What was verbalised by the management was having seized the opportunity to create a space for older Anglo-Indians to call home in their final years. By building their own rest home in Australia, the Anglo-Indian community in Melbourne have therefore publicly and materially claimed the right to be there, and their right to grow old and look after their own in their own way.

On spending more time at St Joseph's, I became aware that the function of the hostel has been expanded to involve more of the local Anglo-Indian population than only the residents and their families who visit. An Anglo-Indian association member described the hostel as 'a sort of focal point' for members of their association, which has close to 1,000 financial members (personal correspondence, July 2009). The association holds its monthly committee meetings at the hostel, and uses the hostel address for correspondence. St Joseph's is frequently the starting point for the bus tours that are run regularly as a fundraiser for projects for the less fortunate of their community in India. Association members call in on bingo afternoons, drop in with gifts of baking and fruit and call by to spend time with residents. Visitors are welcomed and encouraged. The decision to use the hostel in this way contributes to the sense of this being an Anglo-Indian-specific establishment of value to many more than just the 15 residents it houses. The hostel is well known in the wider community, and seems to be viewed with a sense of pride. Melbourne-based Anglo-Indians that I spoke to were always pleased to hear that I was spending time in 'their' hostel.

Local Anglo-Indians who come to the hostel and interact with the residents do so as much to meet their *own* needs, I would suggest, of spending time in a dedicated Anglo-Indian space, as to meet the needs of the residents. In Appadurian terms, 'neighbourhoods are ideally stages for their own self-reproduction' (1996: 191). Not only have they produced this locality, ostensibly for the older members of their community, but through regular contact with the hostel their own 'Anglo-Indianess' is maintained, and perhaps augmented, as they immerse themselves in this taken-for-granted structure of feeling, this locality-producing neighbourhood.

Conclusion

St Joseph's Hostel has been set up so that older, less independent Anglo-Indians can still be Anglo-Indian in their day-to-day practices. The combination of factors including food, the evidence and practice of Christianity, accents and colloquialisms, gave me the sense of being plunged into Anglo-India, and it has the same effect on the residents, even if this is not consciously recognised and identified. The hostel residents are a relatively immobile group who have retreated from wider Australian social spaces into a space that has been created in such a way that they identify with it culturally and where they can feel at home.

Appadurai, in writing about transnational movements, draws attention to the ways in which the 'local' is re-established by transnationals of various types: migrants, refugees and even tourists, with varying degrees of success (Appadurai 1996: 191–4). In Australia, this task is being performed by Anglo-Indians on a number of fronts: with the formation of Anglo-Indian associations in the major cities (with or without the material space of club rooms), the creation of the physical and cultural space in the form of an Anglo-Indian cultural centre in Perth, and most importantly, the establishment and running of the rest home in Melbourne. While the foresight of an Anglo-Indian association made the establishment of this hostel possible, it is also the practices of the residents who fill and transform this space on a daily basis that make this an Anglo-Indian locale. The aspects of material culture, such as the food and Christian iconography, mark their shared identity in ways that enable a sense of familiarity. While Appadurai's theory does not rely on materiality, in this case study, 'locality' would not be produced without the bricks and mortar, the hanging of particular pictures, and specific cuisine, alongside the non-material elements such as entertainment, rituals and sociality. Appadurai's distinction between neighbourhoods and locality thus does not hold in this ethnographic context, as memory, habitual practice, structures of feeling, social networks and materiality all fuse to create a sense of local belonging. Furthermore, despite being aged and relatively immobile, this community's history and its members' life histories are founded on global mobility and movement. Even in their localised and bounded migrant space, the residents utilise the realms of imagination, memory and material ritual to transgress geographic borders. Transnationalism and localisation are not opposing forces in this case study, but are equally integral in the creation of migrant lives.

Symbolically, in having founded a place and made it a local Anglo-Indian institution to cater for the needs of their ageing population, Melbourne's Anglo-Indians have created an important focal point for the wider community. The imaginary of India is still strong for many Anglo-Indians now residing in Australia. They still have what one elderly man described as a 'heart-connection' with India. In St Joseph's, whether as residents or visitors, even with the eucalypts within sight, they can feel Anglo-Indian again.

References

Andrews, R. 2005. *Being Anglo-Indian: Practices and Stories from Calcutta.* Palmerston North: Massey University.

Andrews, R. 2008. A day at St Joseph's Hostel, Melbourne, in *The Way We Are*, edited by L. Lumb and D. Veldhuizen. New Jersey: CTR, 170–77.

Anthony, F. 1969. *Britain's Betrayal in India.* Bombay: Allied Publishers.

Appadurai, A. 1996. *Modernity at Large: Cultural Dimensions of Globalization.* London: University of Minnesota Press.

Bear, L. 1998. *Traveling Modernity: Capitalism, Community and the Nation in the Colonial Governance of the Indian Railways*. Michigan: University of Michigan.

Bland, M. 2004. All the comforts of home? A critical ethnography of residential aged care in New Zealand. *Nursing Praxis in New Zealand*, 20, 57–8.

Blunt, A. 2005. *Domicile and Diaspora: Anglo-Indian Women and the Spatial Politics of Home.* Oxford: Blackwell.

Bönisch-Brednich, B. 2002. *Keeping a Low Profile: An Oral History of German Immigration to New Zealand.* Wellington: Victoria University Press.

Bourdieu, P. and Wacquant, L. 1992. *An Invitation to Reflexive Sociology.* Chicago: University of Chicago Press.

Caplan, L. 1995. 'Life is only abroad, not here': the culture of emigration among Anglo-Indians in Madras. *Immigrants and Minorities*, 14, 26–46.

Chattophyay, D. 2007. Diversity matters: Anglo-Indians: new old-world charm. [Online: Help the Aged e-news]. Available at: http://www.helptheaged.ca/enews/april07/story3.htm [accessed: 10 February 2010].

D'Cruz, G. 2006. *Midnight's Orphans: Anglo-Indians in Post-Colonial Literature.* Bern: Peter Lang.

Glaxo-Smith Kline. 2005. Consumer healthcare history. [Online]. Available at: http://www.gsk-ch.in/History.aspx [accessed: 10 February 2010].

Grunwell, R. 2000. Sacked Chef's food not bland enough, in *Sunday Star Times* (NZ), 6 February, A3.

Hage, G. 2004. *Against Paranoid Nationalism.* Annandale, NSW: Pluto Press.

Jackson, M. (ed.) 1996. *Things as They Are: New Directions in Phenomenological Anthropology.* Bloomington and Indianapolis: Indiana University Press.

Jordens, A. 1995. *Redefining Australians. Immigration, citizenship and national identity*. Sydney: Hale and Iremonger.

Lewin, E. 2005. Perceptions about skin colour and heritage: the experience of Anglo-Indian women in western Australia. *Social Identities*, 11, 631–51.

Myerhoff, B. 1978. *Number Our Days.* New York: Simon and Schuster.

Patel, K. 1990. India, in *I Have in My Arms Both Ways: Stories by Ten Immigrant Women*, edited by A. Jansen. Wellington: Bridget Williams Books, 1–21.

Stoller, P. 1989. *The Taste of Ethnographic Things: The Senses in Anthropology.* Pennsylvania: University of Pennsylvania Press.

Sutton, D. 2001. *Remembrance of Repasts: An Anthropology of Food and Memory.* Oxford: Berg.

Turner, V. 1978. Foreword, in B. Myerhoff 1978, *Number Our Days*. New York: Simon and Shuster, xiii–xvii.

Williams, B. 2002. *Anglo-Indians: Vanishing Remnants of a Bygone Era.* New Jersey: CTR Books.

Younger, C. 1987. *Anglo-Indians: Neglected Children of the Raj.* Delhi: B.R. Publishing.

PART IV
Contesting Urban Place

Chapter 8
Invoking a Community of Engagement: Mobility and Place in a Small English Town

Karen O'Reilly

This chapter draws on ethnographic data from a small East Midlands town in England to demonstrate that local community is a resource for identity formation and expression, as well as collective claims for place-based material resources and emplaced social networks of daily relationality. Community is brought more into play in times of anxiety, such as when urban restructuring, flexible labour markets, globalising processes and increased levels of migration impact on people's daily, lived experiences. The expression of community boundaries, however, is not necessarily drawn foremost along ethnic/foreigner/immigrant lines, but along the lines of mobility and instability. Migration, in the situation I describe, is thus feared not for the immigrants it brings but for the changes it promises, and especially the related transience and loss of continuity. Disputes about the symbolic borders of the community often play out with regard to debates over the real geographic and physical borders of the town, and disputes over the economic and political boundaries and the material resources these enable or disable. Emplaced memories and narratives of the past are central tools through which people recreate community identity and belonging in public spaces on a daily basis and effectively encourage locals (of all backgrounds) to collectively fight for or defend resources. Overwhelmingly these narratives deploy notions of both victimhood and a community of engagement as a counter to the threat of transience.

Real Lives

Mrs Walton is 85 years old. Her husband, who died 10 years ago, was the general practitioner in Shepshed for 30 years. She was introduced to me as the doctor's wife. Mrs Walton lives alone in a big house with five bedrooms; her sons have moved away to London and Birmingham. Although she has dementia, she manages to live alone because she has a gardener and a cleaner who call in regularly, and most days she walks across the road to the chemist or the post office or drives to the Co-op down the road. She also eats out five days a week, in a pub in the next town, with some friends. She never eats at home because, she says, 'I'm never hungry so I forget to eat, but when I go out I feel like eating'. Mrs Walton's sons worry about her but she refuses to be moved out of her home until (in her own words) she is carried out in a coffin.

Jim is 35 years old. He used to work as an electrical engineer but in his spare time he enjoyed driving his Kawasaki motorbike on long trips around the United Kingdom with his wife on the back. But one day a car driver stopped suddenly in front of him and Jim crashed into the back of it, smashing his skull and ribs and breaking his neck. Jim is no longer able to care for himself and now lives in supported housing in Shepshed, 100 miles from his home town, and is not permitted to see his sons because he finds it difficult to control his temper: 'I don't mean it, but I just get angry sometimes and I shout and stuff'.

Tom lost his sight when he was 11 years old. At 26, he still lives with his family in Shepshed and is very happy there. Once a week he makes his way down to the community centre to meet with other blind and partially sighted friends. 'I walk there all by myself', he told me proudly. They make things together, drink coffee and tea, and share stories. Tom makes furniture for doll's houses – tiny tables, chairs and lamps – which he sells to raise money for the community centre.

Tracy is a mother of two young children. She has lived in Shepshed all her life; her mother, father, grandmother and two aunts also live in Shepshed, as do Tracy's sisters and brother. Tracy's husband, Glen, works in the nearby town of Loughborough. It is only 10 miles away but Tracy told me she wished he could find work in Shepshed. Tracy never wants to leave her home town. 'I love it here. I know everyone and everyone knows me.' She has been as far as London once, but she didn't really like it. 'Nottingham is enough of an adventure', she said.

These were some of the many people I met when I began my study of migration in Shepshed. They are illustrative of the arguments I elaborate below – that place-based, material resources, face-to-face communities, localised identities and a sense of shared history and commitment to a geographical place are especially important for some groups of people, especially those who lack mobility or its desire.

Shepshed

Shepshed is a small east Midlands town, with a population of approximately 14,000, centrally located in a region that has experienced dramatic changes in recent decades as a result of immigration from various parts of the globe. Shepshed is 30 miles from Leicester, a large city of 285,000 people which is forecast to be the first city in the UK with a majority non-white population. Leicester has a very large south Asian community, which settled for the main part during the 1970s and 1980s. But migration has been an important feature in the region more widely with, in the past few years, Shepshed and nearby towns experiencing a notable migration of Polish and other eastern European migrants looking for work in agriculture and industry. Nevertheless, Shepshed remains predominantly white (98 per cent in the 2001 census) with some small visible presence of an historical migration in the shops and restaurants owned and run by ethnic minorities and in the businesses around the edge of the town that employ people from a range of backgrounds. The town is mentioned in the Domesday Book[1] but grew especially as a centre for the wool trade during the seventeenth century, and as a result of factory production of hosiery in the twentieth century. Now, however, since the construction of the M1 motorway in the 1960s and the closure of all the factories in the past 20–30 years, it has become known as a dormitory town (a town where people live but don't work) for Loughborough, Leicester and Nottingham. This label is used overtly on the town and borough council websites as well as the website of the Member of Parliament for Loughborough, Andy Reid.

There is a very strong, thriving community in Shepshed. There are primary schools, a high school and community college, two bustling cafés, community centres where a range of societies and groups meet regularly, a thriving library, a (small) Friday market, two post offices and several small shops of various kinds including a butcher and baker. People are seen walking about the town, especially on market day. Kim's Café is a crucial institution, a central meeting place for teenagers, people with learning difficulties, young mums and children, working people popping in for a sandwich, older people taking advantage of special deals and just about everybody else. The chemist on Charnwood Road knows most of his customers (and their ailments and medicines) by sight. In the post office one can overhear neighbours catching up on the latest gossip, families finding out about loved ones they haven't seen in a few days, lonely people sharing a few words with anyone who will listen and young people making arrangements for their social lives. Many people told me how much they love the town, describing the neighbourliness, the good facilities, the range of social clubs and the sense of community. Nevertheless, at the very outset of my ongoing ethnographic study in Shepshed, looking to examine the impacts of migration on a small town, I was overwhelmed with stories of loss. For many Shepshed people (of all ethnicities) the label 'dormitory town' implies migration, rapid change, transience and lack of stability and continuity. Here people are not talking about international but rather internal migration. For them, migration has little to do with race or ethnicity, and has everything to do with community, place, locality and commitment. Surprisingly, in response to my questions about migration to the area, people spoke of their fears that Shepshed is changing, dying, losing its heart (or its core); they especially discuss the loss of material and emotional resources for those who have commitment to, and need, Shepshed as a place.

---

1 The Domesday Book was a land survey commissioned by William the Conqueror in 1085 to assess the extent of the land and resources owned in England at the time, and therefore the extent of the taxes he could raise. The information collected was recorded by hand in two huge books, in the space of around a year (www.domesdaybook.co.uk).

> Shepshed is changing; it is dying. We are becoming nothing more than a dormitory town. There is nothing here for locals, no jobs, no shops, it's all going to Loughborough. (Glen, care worker)

> We are not really aware of the more recent migrations since these people work in the factories near here but tend not to live in the town. (Shirley, library worker)

> … Shepshed changed and it's like you see it now, it's totally soulless, there's no real heart, no civic, there's nothing … The knitting industry sort of drizzled away, but before that things started to change because the motorway came, and I've talked to people who only came to live in Shepshed because it was a good place to get to other places! (Edith, retired)

For many people living in Shepshed the relationship between migration and community is simple: too much mobility is leading to a loss of community.

The Sociology of Mobility and Materiality

Globalisation and increased levels of mobility – of things, ideas, people and capital – have led to a growing concern on the part of social theorists and empirical researchers alike with flows, fluidity, transience and change rather than structures, communities, cultures and peoples (Bauman 2000, 2005, 2007, Urry 2000). There has been an increased emphasis on movement rather than borders, on spaces rather than places. Globalisation, in this rendering, challenges territory, place, nation and state; it threatens borders and controls, and turns its gaze, positively, towards the virtual, the mobile and the communicative (Castells 2001). But, perversely, it seems the constant movement, individualism, competitiveness, increased insecurities, rapid pace of change, anxieties and fragmentation engendered by the postmodern condition lead humans to revive, with an increased sense of urgency, what is essentially a timeless search for community. 'One of the unintended consequences of modern capitalism', Bauman notes, 'is that it has *strengthened the value of place*, arousing a longing for community' (1998: 138, my emphasis). In the face of detachment people yearn for the sharing, the cosiness, warmth and comfort that community promises (Bauman 2001). But, for Bauman, this search is unfortunately doomed to failure because community demands responsibilities and commitment, which are now rejected as antithetical to individual freedom. This is the paradox of our times.

Apparently, we need not fear that community is gone forever. Contemporary mobile life, postmodern theorists contend, is not bereft of community; the sense of belonging and security it provides can now be found in the form of highly individualised communities, based on diverse means of communication (Delanty 2003). Communities are no longer only located in places or cultures, but need to be traced in the form of networks, communication flows and virtual ties centred around individuals (Castells 2001); they are to be found in ways of *expressing* a sense of belonging rather than in firm structures, places or spaces; and they are fluid, open and changeable. However, such virtual or 'personal communities' (Wellman 1999, Spencer and Pahl 2006), though often portrayed as liberating, are themselves fleeting, transient, insecure and 'unable to resist the forces of globalisation' (Delanty 2003: 195). They do not satisfy the desire for communities of place. 'The revival of community today is undoubtedly connected with the crisis of belonging in its relation to *place*' (Delanty 2003: 195, emphasis added). As David Harvey (1990) contends, such postmodernist renderings of community merely overlook the ground reality for many urban dwellers (a reality of poverty, helplessness, fragmentation and marginalisation) and encourages us all to accept the disintegration or separation of community and environment. It is conditions such as economic recession, loss of industry, migration (including internal migration from rural to urban spaces, and later vice versa), and neoliberalism that give rise to a need to revive or sustain communities. The study of diverse mobilities that transcend local, regional and national borders (Urry 2007), denies or overlooks the relevance of place. For the people in my research, community is not simply symbolic (Cohen 1985) or imagined (Anderson 1983); it does not exist sufficiently in the space of networks and flows, and cannot be decoupled from daily interaction (Amit 2002, see Phillipson and Thompson 2008).

Phillipson and Thompson (2008) suggest that there is little articulated sense of community among working class people; they believe it is much more likely that middle class people will express an 'elective belonging'. Yet in Shepshed, where the community's resources are under threat, it is the working classes, the poor and the powerless who feel threatened by an influx of middle class commuters, by the control of a middle class borough council, by transience, change and loss. Here a sense of belonging is invoked that is tied to a commitment, over time, to local and family history. So rather than the 'elective belonging' of the middle classes, there is a belonging based on investment, an *engaged belonging* which deems itself worthy as a result of history, time, locality, work, family and commitment. This is therefore a community opposed to mobility and movement, transience and change, since these do not allow for such investment and engagement. Without investment there is the threat that the community will die. However, most interestingly, some ethnic minority groups and some recent newcomers are able to claim membership of the community as a result of expressions or acts of investment (of time, effort and concern) and so the community is not opposed to ethnic minorities so much as to mobility and transience. As Vered Amit (2002) suggests, community is first and foremost a claim of, or for, social engagement.

Material Culture

People render their communities real to outsiders (and participant observers) through the stories they tell and through the public face they reveal via events, actions, leaflets, books and photographs. Certain events, discussions and interviews in my research are notable for the centrality of material factors; locals frame their belonging in terms of the loss or struggles over parks, buildings, shops, markets, banks, housing estates, the swimming pool, the volunteer service, the housing estate and the motorway. Kathy Burrell (2008: 64) found the same in some of her interviews with Polish migrants. This is not surprising, she says; others have noted 'the emotional connection between people and things' (see Miller 2005). In Shepshed, certain events and places have become synonymous with, or symbolic of, continuities and others with change; certain things are imbued with memory, others represent loss. Interviews highlight how people's economic and cultural experiences are specific to material contexts, and to move away from (or to lose) the material, physical context implies losing one's past experiences and one's security. As Morton (2007: 158) has suggested, the interweaving of stories of the community with memories of the past and fears about the future built environment are 'suggestive of a deeper relationship between dwelling and social life'. The built environment clearly influences the way people think about and perceive their world, but more than that the material provides or denies real resources needed by people in their daily lives. Morton talks of 'memory work as a social process related to the house' (2007: 160). He investigates the ways in which 'the material home may structure and influence the reproduction of embodied forms of memory'. If what we build in the future is a way of collectively (re)constructing a memory of how things were in the past, then having no control over this is anxiety-ridden and threatens that we lose our memory of who we were as well as actually losing our community and its resources in the future.

The Material Loss of Community

Overwhelmingly, what is being expressed in Shepshed is a fear of the loss of sufficient resources to sustain an emplaced community in the future. Community is thus expressed in terms of place-based material resources, such as shops, jobs and housing, and it is claimed within an emplaced, embodied and intricately networked being and belonging. I have been told numerous stories of resources that are already lost or under threat of closure, and of goods and services that are no longer locally available. A major, overtly expressed anxiety was with the number of shops that are closing down. This is a more pressing fear in the current economic climate but is not a new concern. If there are too few shops, it is feared, few locals will spend time in the town at the weekends or during the daytime. This will leave the place empty, with no real heart or centre, leading to increased crime and run-down areas. This is not a baseless fear: shops in the town were already closing before the recent recession and at a last count nine shops and businesses in the town centre were shut and four were boarded up. The town centre looks scruffy in some parts and deserted in others. During one walk into town on a Friday (market day in Shepshed), I got talking to an elderly man who told me, as we walked along the road side-by-side, that his wife had died a couple of weeks ago, just before their 52nd wedding anniversary. He was clearly distraught and looking for anyone to talk to about his loneliness. I had seen him leave the tattooist as I walked past and have a quick chat with another woman before crossing the road to me. He told me he was dreading going home to his empty house, and then he said: 'But Shepshed isn't like it used to be. You don't see people you know walking about like you used to, because there are hardly any shops left now'. Shops are thus representative of a sense of locality, a place-based material and active, embodied belonging to the town. Without shops there is less reason to spend time in the town.

The loss of local jobs too, is shown to represent a loss of emplaced community. As Manuel Castells (2001) notes, local communities are not so much undermined because of mobility (which he believes has been exaggerated) but because of the changing nature of work. It was indeed work that kept people in contact within their villages and towns and formed the basis of their interaction, not just its existence but the rhythm of work and the regularised pattern of a working week. As Edith, a retired woman who has lived in Shepshed all her life, told me in an interview:

> We used to have loads of shops, and when the factories closed on a Friday night all the ladies would rush out with their pay packets – I've got me money – and the town buzzed, absolutely buzzed, because you went and got your groceries, you went and got your meat, you paid your papers and everybody was there rushing around … As the factories closed so did the shops start to close.

People worry about the lack of jobs in the area because this takes people out of the town every day to find work elsewhere and thus undermines the intricate networks of emplaced relationality on which their daily lives were based. But a lack of local jobs can also lead to high rates of unemployment if enough local people are not able to compete in the wider labour market, and people are justifiably worried about the social ills that high unemployment can mean for an area. Again, these fears are not ungrounded. In many interviews and conversations locals recalled Shepshed's past as a centre for hosiery; the prior vibrancy of the town has been described to me at length and its employment history is celebrated in a range of publications about the town. But this social history is now a story of loss, of closure and of rapid change.

> In the 27 years I've lived here I've seen a lot of changes, ever such a lot. Oh yes. There's been lots come to live here for work first, but now they live here to go elsewhere. (Carolyn, office worker, from field notes[2])

> Shepshed has changed beyond all recognition: it was very much a village, people knew each other … It always had a lot of hosiery factories, well the hosiery

2 Quotes 'from field notes' are not transcribed from recorded interviews.

> industry began in people's own homes and didn't go into factories until later on, and these were very small factories ... People worked in hosiery all their lives, the men worked the machines, and there was a great tradition of outwork – the women working in their own homes doing mending and linking, and framework knitting was sometimes done where people had their homes adapted to fit these machines in. (Freda, retired)

Locals also worry that there will not be enough reasonably priced housing for their children to buy when they grow up. Children will thus be forced to move away, whether they wish to or not, which in turn will mean locals do not see so much of their children and grandchildren, are compelled to travel more often themselves and have no one left to look after them as they age. This is contradicted by the argument that there is too much cheap housing, which runs its own risks of attracting, as Bernard, a retired man told me, the 'wrong sort of person to the town. I mean we all know that people who get cheap homes don't invest so much in looking after them, don't we?'

Behind these key concerns about jobs, housing and shops is a sense the town is changing against people's will; and the tales of loss can be supported with material evidence. The town still has two post offices, but only because locals held a fierce campaign to fight closures. There is a very busy volunteer centre, but the organisation that runs it has now merged three small units into one with the result that the town's local office is only open in the mornings and will probably close altogether in the near future. There are only two ATM machines in the town now, and the only bank recently shut despite a protest against its closure. There has been a campaign to improve and maintain the public toilets in the town but these remain a disgrace, while Loughborough's public toilets have won awards. There used to be a Christmas parade with a Santa on a sleigh, but this has been cancelled because of difficulties of gaining adequate insurance against risk. There is an ongoing campaign not to have the M1 motorway (which runs past Shepshed between the town and Loughborough) widened to three lanes. There is no longer a carnival because the organisation that ran it has folded. There was a community college running a wide range of adult education classes but this has closed down. Expressing his own concerns about the undermining of place-based communities, a pharmacist told me:

> The government is allowing chemists to merge into bigger and bigger concerns where customers are no more than a number – no one knows anyone any more. It used to be that my job involved talking to people and giving a personal service. I know the doctors at the surgery, they can ring me if they have a question about someone's medication and I can ring them. (Sam, pharmacist, from field notes)

There is an overwhelming feeling that the ongoing loss of both material resources and of local, face-to-face networked relationality is occurring against the will of those it affects most, leading to a sense of both struggle and defeat.

Loss of Control Over their Own Fate

Shepshed town is part of Charnwood Borough, in the County of Leicestershire. It is very difficult to work out which council (town, borough or county) has control over which resources. But Loughborough, a town with a population of approximately 50,000 and very close to Shepshed geographically (separated only in material terms by the M1 motorway and the Garendon estate) is clearly seen as a threat to Shepshed's identity and political and economic autonomy. A key difficulty is that Loughborough is the largest town in the borough, with a third of the borough's population. Although still relatively large, Shepshed is small in comparison and therefore has to compete for town resources. Many Shepshed locals would like their town still to be a village and often use that term when describing it. The greatest fear they have is that it will be merged, physically and symbolically, with Loughborough as part of the neoliberal restructuring affecting small towns across the country. While they demand the resources that a town of its size needs, they feel that most resources go to Loughborough as the bigger of the two towns. Other villages in the borough have managed to retain their identity as villages, as rural and middle class. Shepshed has an industrial past, is more working class, and locals express a feeling of having lost control over its future. At a council meeting, for example, one of the councillors was grumbling that they had written to the borough council (based in Loughborough) about the naming of a street and no one had replied to her letters. 'They just ignore us. We cannot get a response', she said. Another councillor told me:

> Since the urban district councils joined with the others in Charnwood Borough they feel it is all about Loughborough. The trouble is many of the councillors don't live in Loughborough but it feels as if it's all they care about. There are 150,000 people in Charnwood and over 100,000 do not live in Loughborough, yet all our taxes are spent improving Loughborough. (Anonymous, from field notes)

Loughborough also represents a threat to the town because, as a result of urban restructuring and economic and political centralisation, it has the material resources to attract people from Shepshed for shopping, banking, eating out and social activities.

> Until a few years ago Shepshed was a village and it was self-contained. We had all the shops we needed and we didn't have to go into Loughborough. Now some people blame Loughborough for what they have lost, their community college, their shops. The community college used to have all sorts of evening classes but now there isn't enough money, these things are no longer funded locally, and there is so much emphasis on learning and improvement nowadays. (Fred, volunteer worker, from field notes)

Even a study undertaken to try to understand Shepshed's needs for the future was organised by the borough rather than the local town council, and 'totally missed the point' (Bruce, town councillor) and 'was a waste of time and money' (Mary, library worker). A sense of powerlessness and frustration pervades many stories. Shepshed people believe they are seen as lower class, or as poorer, or rougher. It was described to me as a 'dumping ground', with lots of care homes, social housing and temporary homes for those waiting to be permanently housed, rather than the facilities they really need.

> They dump their baddies here, from Sileby across to Loughborough, they dump them in those council-owned flats, they dump the druggies and everything. (Carolyn, office worker)

> Loughborough always treats us as if we are a bit common. Woolly backs, that's what they used to call us. It's always been looked down on, it's always been a bit scruffy, a bit down-market. (Edith, retired)

The people who express a fear that Shepshed is losing its heart, or is dying, are people who need local material and emotional resources, be it friendship, employment, a home, shops, a bank, a social or community centre, local services, familiarity, a church, a council with its own resources or somewhere their children can live when they grow up. Men and women working as volunteers in the town, councillors of all political parties, a vicar, a schoolboy, a newsagent, a chemist, a doctor and a schoolteacher, along with many others, started our conversations with phrases such as: 'No one seems to care about Shepshed any more' (Denise, town councillor), 'Shepshed has died in recent years' (Ruth, retired), 'Changes have been made to the detriment of Shepshed itself by outside sources' (George, town councillor), 'It needs a heart, a core' (Julie, full-time mother), and 'It used to be that people grew up and stayed here, but they can't now. It's just a dormitory town, no work' (Dan, barman). Some people, especially those outlined at the start of this chapter, really need these resources locally. Others simply have an identity as Shepshed folk and want to enjoy a local identity and be part of local networks.

The Threat of Migration or Mobility

The fears and concerns expressed above are often framed in terms of lack of commitment to, or investment in, the town. Community is thus something within which resources and identity are mutually constituted. People particularly worry that too many of those who live in the town simply use it to sleep in, working and shopping and socialising elsewhere. In other words, they worry about a lack of emotional investment as well as physical presence. The phrase 'Shepshed is just a dormitory town' (Brenda, care worker) is a negative denigration of mobility, not a celebration. In their important new book about the social cost of inequalities, *Spirit*

Level, Richard Wilkinson and Kate Pickett (2009) observed that when people do not see each other on a day-to-day basis and there is little face-to-face contact in the community, there is decreased trust, which in turn leads eventually to ill health. Again material factors are brought into play. The housing estates that have been built more recently on the edge of the town provide accommodation for outsiders (in most cases, migrants from other parts of the UK). The M1 motorway, though it has some positive symbolic function in separating Shepshed from Loughborough, makes it easier for people to live in the town and work elsewhere. The migration that is feared is the movement into the town of people who constantly move out of it, every day for work, at weekends for shopping or permanently at some point in the future. Barbara, a full-time mother, described to me the tremendous changes she had seen just in her 27 years in the town, with the growth in numbers of residents working in the factories, the factories closing down and then a new influx of people moving to the housing estates.

> But a lot of people move here because of the motorway. It's known as a dormitory town ... it's been proven that a lot of people on that new estate (pointing) don't come into the town at all. They don't even know there's a Co-op.

And others grumbled:

> Shepshed went wrong somewhere. It had the potential to be a nice little market town. I think the problem is that a lot of people live in Shepshed but they don't shop here. They drive out of Shepshed. We're a dormitory town really. People always move on, move out, that's the trouble. (Ruth, retired)

> There are too many new people now, like students, who just stay a while and then move on. (Woman at a market stall, from field notes)

Students, too, are seen as part of the dormitory phenomenon. Like ethnic groups in the past, they are treated as the dangerous outsiders, bringing crime, unsociability, lack of commitment to the town and transience. These, and the dormitory residents, are the immigrants that Shepshed fears. People in Shepshed, interestingly, do not seem to fear migrants so much as their migration. There is some evidence that the British National Party (BNP) is attracting more support than previously, which is usually an indication that racism or xenophobia are on the increase. But most people I spoke to were more concerned about the rise of the BNP than they were about immigrants themselves. Shepshed folk worry that mobility, rather than immigration, will lead to a loss of identity for the town, to fewer people caring about it as a locality and fighting to maintain it in the face of other changes.

Celebrating an Engaged Community

Community identity is a way of talking about 'us', and 'us' is a collectivity that can help fight for resources. In Shepshed this collectivity, the community that is expressed or invoked, includes Asian shopkeepers, an Afro-Caribbean minister, a vicar originally from Singapore, a Greek family, as well as Shepshed folk 'born and bred' (people who have not only spent their whole lives in the town but can claim a relationship going back generations). As June, a volunteer worker, told me, 'Shepshed has a strong sense of community, and ethnic minorities are now part of that community'. Community is expressed here in terms of history and commitment; it is an *engaged community*. Anyone who can invest or has invested time and energy in the community is worthy of being a community member.

Community membership is invoked most strongly through historical ties. People are considered, or see themselves, as members of the community because they can talk about the past, how long they and their families have lived in the town, thus demonstrating not only identity but also a temporal and emotional investment in community. People therefore talk about their past working lives in the town, or describe their family histories; they invoke memories of how things were, as seen in many quotes above, articulating an identity with place and locality. Others claim membership through virtue of the fact they not only live in Shepshed, but shop there, use the chemist, the supermarket and (importantly) do not rely much on the things Loughborough offers. Others talk about their community involvement, voluntary work, council membership, community groups they belong to, campaigns they have joined and so on. People also invoke community by telling me of others in the town I should talk to: Hilda who is 96 and lived here many years; the Asian chemist who knows everyone; the Indian man who owns the hardware shop and can get anything you need; the Greek family that owns several small shops on 'the parade'; Dr Singh, the general practitioner who has worked there for 30 years; the liberal councillor who has made it her mission to get a second supermarket in the town; the teacher at the community college who 'doesn't even live in the town but does a lot for it'. In all these invocations it is time, history, commitment, investment and engagement that matter; in other words the opposite of all the things that mobility calls to mind.

Many of the themes discussed above are illustrated in the ongoing campaign to block plans to build 3,500 new homes on a piece of land between Shepshed and Loughborough known as Garendon estate. Garendon is the estate of a very large old house that burned down in the late 1950s. The owner of the estate, the Squire of Garendon, also owns much of the land in Shepshed. Garendon Park, part of the estate, is protected as a national heritage site; under these plans the park will remain intact but be surrounded by urban dwellings. Garendon Park and estate hold important symbolism and materiality for Shepshed people. The estate is used for walking and cycling and has a path running through it that links the town to Loughborough; but it also symbolises the separation of the two towns. It is also part of Shepshed's social history; as Edith told me, 'There's a bit of a folk memory of when Garendon was the big house, and your parents and grandparents worked there, and it was very important to some families'.

In protest to the plans, the Garendon Park Countryside Protection Group (GPCPG) has its own newsletter and a dedicated website. A series of meetings have been arranged (by both the council and the protestors) to which local people have been invited to discuss the plans. Over 100 locals turned up at the Town Hall when the Borough Council attempted to justify Garendon as its preferred location for the development. Shepshed people were clearly not happy with the plans. They objected that the development would render their town indistinguishable from Loughborough, except for the M1 motorway; they want to be separated by green space, not a motorway. They grumbled that the development would not be of any benefit for Shepshed, since the inhabitants of the new homes would work and play elsewhere. Concerns were raised that Shepshed and Loughborough would simply merge into a huge urban area without any extra services or jobs. Some protested about the extension of what already amounted to their treatment as no more than a dormitory town; others felt the area would become even more run down and abandoned. It became very clear what people were fighting to retain: their own 'village' identity, local facilities, a green surrounding area, local jobs and low crime rates. Sue, a teacher, told me of Garendon estate:

> It's the green wedge between us and Loughborough and we don't want that lot here do we? … It's creeping closer all the time. There was a time when we were very separate. In fact a lot of people were worried about the M1 coming this way, but in fact it stopped a lot of people coming to Shepshed, it's a bit like a motor between us and them. We don't want to be built up, we want it open and green … Eventually, if we are not very careful, I think perhaps eventually we are going to be tagged on to the end of Loughborough, and then we'll be the dog-end of everything won't we?

The debate about the development of Garendon Park estate serves as a reminder of the town's history. It acts as a reclaiming of that piece of land as Shepshed's, a collective remembering of a shared history, as well as an opportunity to express a unified identity. The resistance mirrors the defence of their locality against the wider world, against change and, above all, against mobility and transience. The event at the Town Hall was thus a ritual expressing a symbolic boundary between Loughborough and Shepshed, which served to 'heighten people's awareness of and sensitivity to their community' (Cohen 1985: 50); this kind of ritual is often to be witnessed when the geo-social boundaries of a community are undermined or threatened.

Conclusion

Shepshed is a small town that has experienced rapid and drastic social change in the past few decades. Factories have closed down and locals now have to travel beyond their local community for work. Shops and pubs have closed, younger people are moving out for work elsewhere and little is left of what was once a thriving hub. Meanwhile, a motorway now passes so close to the town that it is possible to travel to nearby cities in short periods of time, and housing estates have been built to attract newcomers (migrants) who have no commitment to, and little time to spend in, their local community. Those who are left behind, those who need a local community for the material resources as well as the company it provides, have real fears that those things which represent community, security, locality and constancy are being eroded at a pace of change over which they have no control. They can see their town becoming no more than a neglected part of the urban sprawl, tagged onto a larger town as an afterthought, a dormitory town where residents spend their nights each in their own home, to depart every day in a constant process of dwelling and travelling. In response to these issues, local people in Shepshed are not expressing fears about ethnic minorities or immigrants; what they fear is mobility itself. What they celebrate, then, is an engaged community – an inclusiveness based on time, history, commitment and investment – because only those who engage with their communities are prepared to fight for them.

It seems the constant movement, individualism, competitiveness, increased insecurities, rapid pace of change, anxieties and fragmentation engendered by the postmodern condition lead humans to revive, with an increased sense of urgency, what is essentially a timeless search for community. The virtual or communication-based personal communities of mobile individuals cannot satisfy the need for place-based, local, face-to-face contact, for some groups especially. The newly mobile world in which we live seems to have no place for the immobile, those for whom movement is difficult because of physical or financial limitations, or those who have insufficient resources with which to join the network society. As a result, mobility itself is feared and we see a revival of 'engaged communities' invoking history, commitment, time and, perhaps, stasis.

References

Amit, V. 2002. *Realizing Community: Concepts, Social Relationships and Sentiments*. London: Routledge.

Anderson, B. 1983. *Imagined Communities: Reflections on the Origins and Spread of Nationalism.* London: Verso.

Bauman, Z. 1998. *Globalization: The Human Consequences*. Cambridge: Polity.

Bauman, Z. 2000. *Liquid Modernity*. Cambridge: Polity.

Bauman, Z. 2001. *Community*. Cambridge: Polity.

Bauman, Z. 2005. *Liquid Life*. Cambridge: Polity.

Bauman, Z. 2007. *Liquid Times. Living in an Age of Uncertainty*. Cambridge: Polity.
Burrell, K. 2008. Managing, learning and sending. The material lives and journeys of Polish women in Britain. *Journal of Material Culture*, 13(1), 63–83.
Castells, M. 2001. *The Internet Galaxy, Reflections on the Internet, Business and Society*. Oxford University Press: Oxford.
Cohen, A. 1985. *The Symbolic Construction of Community.* London: Routledge.
Crow, G. 2008. Recent rural community studies. *International Journal of Social Research Methodology*, 11(2), 131–9.
Delanty, G. 2003. *Community*. London: Routledge.
Harvey, D. 1990. *The Condition of Postmodernity: An Inquiry into the Conditions of Social Change*. Oxford: Blackwell.
Miller, D. (ed.) 2005. *Materiality (Politics, History and Culture)*. Durham: Duke University Press.
Morton, C. 2007. Remembering the house: memory and materiality in northern Botswana. *Journal of Material Culture*, 12(2), 157–79.
Phillipson, C. and Thompson, P. 2008. Whither community studies? A special issue on researching community studies, past and present. *International Journal of Social Research Methodology*, 11(2), 87–91.
Reid, A. 2009. Member of Parliament for Loughborough Constituency, Shepshed Page. Available at: www.andyreedmp.org.uk/shepshed.html [accessed: 26 September 2009].
Spencer, L. and Pahl, R. 2006. *Rethinking Friendship: Hidden Solidarities Today*. New Jersey: Princeton.
Wellman, B. 1999. The network community, in *Networks in the Global Village: Life in Contemporary Communities*, edited by B. Wellman. Boulder, CO: Westview, 1–48.
Wilkinson, R. and Pickett, K. 2009. *The Spirit Level: Why More Equal Societies Almost Always Do Better.* London: Allen Lane.
Urry, J. 2000. *Sociology Beyond Societies. Mobilities for the Twenty-First Century*. London: Routledge.
Urry, J. 2007. *Mobilities*. Cambridge: Polity.

Chapter 9

Negotiating Religious Expression and Citizenry Belonging: Bosnian Experiences in Suburban Melbourne

Lejla Voloder

Fields of belonging are, as Olwig (2002: 124) suggests, cultural constructs, imagined and sustained by individuals and collectives in numerous ways. For migrants, these fields may capture the local through to transnational connections, evoking and mobilising loyalty to different communities simultaneously (Lovell 1998: 5). Thus, despite possible contemplations of returning 'home', and actively fostering transnational connections and diasporic imaginings (which might be suggestive of experiences of 'displacement' and 'dislocation'), migrants continue to become involved in and interact with various social institutions, networks and discourses of the society in which they reside. It is through these processes of engaging in various realms of society that they establish fields of belonging within their local contexts and thus become emplaced. The local, however, is not a bounded cultural site, but rather reveals itself as a juncture of intersecting ideas, values and loyalties (Gupta and Ferguson 1997, Olwig and Hastrup 1997). In this way, belonging in and to the local is a process of dialectical engagement with multiple and sometimes contradictory discourses and practices of inclusion. The determinants of belonging can be many and varied. Debates over the parameters of national belonging and citizenship can be instrumental in affecting modalities of belonging in local spaces. The prominence of transnational studies in recent years has brought into question the salience of the nation-state in mobilising belonging. While connections that traverse the boundaries of nation-states have highlighted the selective permeability of state borders, I agree with Ong (1999: 15) who argues for the need to recognise the continuing presence of the nation-state as a key figure in debates over belonging, particularly as it continues to 'define, discipline, control, and regulate all kinds of populations'. The nation-state consequently imparts and informs the development of various fields of belonging. State involvement in the maintenance of, not only the material but also symbolic, boundaries of citizenship is a case in point.

The substantive dimensions of citizenship, however, are often experienced within localised settings. It is through encounters in the streets, neighbourhoods, workplaces – sites of mundane, daily encounters – that one's claim to citizenry inclusion is shaped, challenged and negotiated. How this occurs in the lives of migrants from Bosnia and Herzegovina[1] in Melbourne is the focus of this chapter. The discussion is based on fieldwork data gathered as part of my PhD research into Bosnian experiences of belonging in Australia conducted in 2006 and 2007. The discussion in this chapter centres on Bosnian Muslim experiences of citizenship within the context of highly politicised discourses surrounding the position of Islam within Western states. In this chapter I draw on the experiences of several key informants to illustrate how migrants assess and negotiate claims to belonging in local spaces.

To explore the way belonging is configured by migrants, I begin by sketching the dominant discourses which frame contemporary debates of citizenship in Australia. The efficacy of multiculturalism in creating and promoting national belonging is one discourse migrants undoubtedly encounter. In addition, in recent years, the debate over the place of religion, particularly Islam, within the Australian national multicultural and secular frame has been potentially polarising, and particularly affronting to Muslim migrants and their attempts to identify with the nation. After situating these broader debates, I will then move to consider how migrants engage with these debates in their local settings. I argue that by engaging with these discourses, migrants seek to locate their own position and identify the possibilities for their own inclusion as citizens.

Interpretations and critiques of the parameters of belonging availed through Australian secular multiculturalism inform the ways in which Bosnian migrants claim belonging in their local settings. Their own assessments of the 'acceptability' of ethnic and religious diversity consistent with 'full' citizenship status frame expectations and enactments of migrants' own religious expression and their attempts to negotiate forms of belonging. Claims to belonging entail attempts to reconcile contradictory interpretations of Australian identification, such as those which speak to a multicultural secular society in contrast to those which promote a vision of a monocultural Christian nation. These discourses have direct implications on the ways in which Bosnians negotiate their own religious expression and how they define, practise and challenge their own belonging in Australia within localised settings.

Dimensions of Citizenship

The experiences of citizenship can be defined in terms of their formal and quotidian aspects. The acquisition of formal citizenship defines one's membership to the nation and marks the legal-juridical relationship between the citizen and the state in terms of defined sets of rights and obligations. However, the experiences of citizenship

1 For reasons of expediency, throughout this chapter, 'Bosnia' is used interchangeably with 'Bosnia and Herzegovina'. In addition, I use 'Bosnian' as short-hand to refer to the migrants from Bosnia and Herzegovina in Australia as this is the term used by my informants.

extend beyond these legal dimensions, as being a citizen is not simply a matter of being accorded the associated rights, but also relates to the manner in which such rights are experienced. Furthermore, formal acquisition of citizenship does not easily translate into the cultural and civic aspects of belonging and identification with the national community. Differential experiences of citizenship may result from the 'informal' subordination of citizens according to race, ethnicity, class and gender (Ong 2003, Rosaldo 2003, Yuval-Davis et al. 2005) and as demonstrated in this chapter, religion. Thus, citizenry belonging is not only articulated in relation to the legal context but in terms of the imagination and maintenance of shared values that are practised in the course of people's everyday lives, in the mundane and local (Ong 2003). And it is often multiculturalism and secularism that are mobilised as values which unite the citizenry in Australia.

Citizenship in Australia is dominated by the multicultural paradigm in both formal and everyday terms. The multicultural model, in formal terms, conceptualises the nation as a united political community. This, as Castles and Miller (2003: 44) contend, allows for the possibility of admitting newcomers to the nation: since membership is defined in terms of sharing a constitution, laws and citizenship, this enables migrants to maintain cultural difference while becoming part of the nation, as long as they adhere to certain principles such as respect for basic institutions and democratic values. Multicultural citizenship thus attempts to reconceptualise migrant belonging in Australia by enabling people of diverse ethnicities to stake their claim in Australian society.

However, within the Australian multicultural framework, the position of religion has been largely ambiguous. This is because the ideology of ethnicity is central to multiculturalism. As McAuliffe (2007) notes, religion and religious identities are largely understood as being outside of multicultural discourse, because 'national' (ethnic) communal belonging dominates the multicultural ideology. Consequently, this marginalises other forms of identity, such as religious identities, from official discourse. The result has been to 'ethnicise' religion, that is, to view religious belonging as commensurate with ethnic belonging (Stratton 1996), and in the process valorise ethnic difference over religious difference and contribute to the normative understanding of multiculturalism as a secular practice which enables national belonging to all religious groups.

In this way, the multicultural frame positions religion as secondary to ethnic identification within Australian society, fostering multicultural belonging through secular ideology. And indeed this notion is supported by the fact that Australia is constitutionally a secular state. In formal terms, secularism has relegated religion to non-political spheres. However, the state of Australia's secularity is in fact much more complex as, in recent years, tensions over the place of religion in demarcating the boundaries of Australian belonging have become particularly pronounced.

Increasingly, the place of religion within the parameters of Australian belonging has been the subject of intense debate. It is the position of Islam and Muslims within the national frame that has been the particular focus of these polemics in Australia and in the West more generally. In the post-9/11 and Bali and London bombings context, questions over the role of religion in Australian identity and national belonging gained heightened significance. Islam in particular has been positioned as problematic within the imaginary of a secular nation and society (Modood 2007, Triandafyllidou et al. 2006). Dominant political discourse has turned toward scrutinising the loyalty of Muslim migrants in particular, questioning both the 'ability' and willingness of Muslims to integrate and belong to Australia. This sentiment positions a particular religion and its adherents as potentially problematic in claiming full citizenship status, thus including religion into the debate over the parameters of national belonging. This discourse has permeated life in the suburbs and the effect has been increases in experiences of vilification and discrimination amongst Australians of Arab-speaking and Muslim backgrounds (Poynting and Noble 2004: 16). Within this context, Muslim migrants' claims to full citizenship have become problematised as they experience anxiety about their place in Australian society.

These divergent political and public discourses inform the ways in which migrants imagine and sustain fields of belonging in the Australian nation, and it is here that the parameters of inclusion promoted within the multicultural frame are contrasted with those which position Islam as problematic for national belonging. By engaging with these conflicting messages, migrants become involved in embracing, challenging and negotiating imposed discourses and ideologies in seeking to identify the possibilities for belonging. In this way, the intentionality in seeking belonging is part of a process, as Ong (2003: xiv) states, of 'figuring out the rules for coping, navigating, and surviving the streets' and everyday circumstances of being a citizen.

I found that for the Bosnian migrants in my study, their experiences included reconciling the proclamations of citizenship promoted by the political and public elite with the ontological dimensions of being a citizen. Within the context of their daily lives, Bosnian migrants engage with these debates by interpreting the possibilities for belonging in a secular nation and how this impacts on public displays of religious expression. In this process, they simultaneously critique the status of Australia's claim to secularity while also debating public displays of religion through which they relay their understanding of what is, or should be, 'appropriate' religious expression in their local settings. Thus, engagement with dominant discourses has direct implications on the ways in which Bosnians negotiate their own religious expression and how they define, practise and challenge their own belonging in Australia within the context of their daily interactions.

Local Encounters with Religious Diversity

Encounters with religious diversity were common experiences for the Bosnian migrants with whom I conducted research. The suburbs of Melbourne in which they settled are sites comprising high migrant populations with greater representations of non-Christian religions than in the broader Melbourne scene. For example, in one of the key suburbs of my fieldwork, 15 per cent of residents identified with Buddhism, 8 per cent with Islam and 2 per cent were Hindu. The diversity of the local population was evident in the streetscape, as in a cross section of streets within walking distance there was an Anglican church, a Chinese Christian church, a Bosnian mosque, a Sri Lankan Buddhist temple and a Vietnamese Buddhist temple.

This diversity, represented by the presence of various places of worship and the religious insignia people displayed in their daily encounters, worked to foster a sense of mutual belonging for citizens. Here, freedom of religious observance was a daily reality in local public spaces, consistent with the civic values of multicultural secular belonging for all Australian citizens. Bosnians accounted for approximately 2 per cent of the local population and were accordingly represented with the establishment of various sporting and social clubs (including choir, youth and pensioner groups) and welfare organisations. The mosque, as the key place of worship, also provided the venue for significant rituals such as weddings, funerals and other commemorative events. Thus, the mosque served not only as a site for the religious but also cultural events – providing an avenue through which an ethnic Bosnian Muslim identification could be claimed within the Australian multicultural frame.

As discussed earlier, the Australian multicultural framework privileges the ethnic identification of citizens, hence contributing to the ethnicisation of religious identification. The result is that ethno-religious identification becomes the basis for belonging to a secular multicultural nation. While this is experienced positively in some circumstances, in others it can be problematic as I demonstrate later. However, first, through a discussion of Mirsada's[2] experiences, I explore how secular ethno-religious diversity is enacted and how it fosters citizenry inclusion.

Mirsada was one of my key informants. She had migrated to Australia in the early 1990s and over the past 12 years her everyday reality entailed encounters with various forms of religious expression. This was reflected in the composition of neighbours with whom Mirsada interacted. She would drink coffee with the Greek-Orthodox widow, Sofia, who lived next door. Drinking 'Turkish coffee', as they referred to it, was a way of sharing moments in each other's lives and recognising the intersections in their cultural traditions as both women had consumed forms of 'Turkish coffee' before migrating to Australia – Sofia in Greece and Mirsada in Bosnia. Talk of religion was limited to discussions of practices enacted on significant events – rites of passage – such as births, marriages and deaths and key religious festivals. Ideology did not enter their conversations. Instead, the sharing of religious practices was captured in the exchange of food. On Easter, Sofia would explain Greek-Orthodox celebrations, treating Mirsada with a plate of 'koulourakia' (butter biscuits) and boiled eggs. On Eid, Mirsada would invite the old woman to her place for coffee and baklava (layered walnut and pastry

2 I employ pseudonyms throughout this paper to honour the confidentiality of my research participants.

sweets), and share the Bosnian Muslim way of celebrating the religious festival. In this way, religious practice was often closely tied in with cultural identity and acknowledging each other's celebrations was a way of recognising difference whilst creating a sense of unity. Indeed, these acts formed part of Mirsada's sense of belonging as they reaffirmed the ways through which both local and national unity was created in Australia.

While such intimate exchanges of food did not extend to all neighbours, Mirsada was aware of the various faiths represented in her Melbourne suburb. She noted the 'Indian Hindu family, the young Vietnamese Buddhist couple, and the Afghan Zoroastrian and his Iranian Baha'i girlfriend'. Her acknowledgment and even positive response to this diversity revealed Mirsada's interpretation that varied forms of religious expression and identification were consistent with claiming belonging to a multicultural secular society.

Negotiating Religious Identification

However, the parameters for expressing this diversity were constrained by other factors and the following account is one example which demonstrates the ways in which religious expression was debated and how practices were negotiated as part of the process of actively claiming belonging. In this snippet from my fieldwork diary from July 2006, I joined Mirsada for an event that was being held in her local community. The entry begins with Mirsada and myself in the car park of a Bosnian Islamic Centre in Melbourne as we prepare to enter the mosque.

> 'Does it look OK?' Mirsada asked, as she tried to manoeuvre the 3m long scarf around her head. I didn't reply. I was caught up in my own struggle, trying to subdue both my hair and scarf, which were flailing recklessly in the strong southerly wind. It was the eleven year anniversary of the Srebrenica genocide and a commemorative event had been organised at the local Bosnian Islamic Centre. We were standing beside Mirsada's car in the car park trying to arrange our headscarves before entering the mosque. Since neither of us wore headscarves ordinarily we were experiencing a bit of trouble, but we knew that, as women, to enter with our hair uncovered was not an option. While we stood trying to catch glimpses of ourselves in the reflection of the car windows, another car pulled up. Mirsada's cousin stepped out, laughing, a witness to our episode of ineptitude. 'Why didn't you get ready at home, you know, like normal people?' Sena mused. 'Oh, I don't know. I didn't want my neighbours to see,' Mirsada replied, gesturing to her headscarf.
>
> Later that evening, back at Mirsada's home as we sat in the lounge trying to fluff up our now flattened hair, a news bulletin started. I had been visiting Mirsada for a few months now and knew that news watching was a particularly active affair in her household. It was not only about *watching* TV, but about *reading*

> it. On this edition, the newsreader was an attractive woman in her 30s. She had shoulder-length blonde hair and wore a green blouse. The top two buttons of her blouse were undone, revealing the base of her neck; her white skin providing a contrasting background to the silver and amethyst pendant she wore on her necklace. 'Look!' exclaimed Mirsada. Sedin, her brother-in-law, obviously on the same wavelength automatically replied in a sarcastic tone, 'Hmmp, and that's appropriate for a newsreader?'. They referred to the cross on the newsreader's necklace. Just then Jasmina, Mirsada's sister, walked into the lounge. 'Look at *that*', Sedin said, nodding at the TV, purposely not verbalising what *that* was. Jasmina, seemingly indifferent, glanced only quickly and quipped, 'So what? It's *their* country'.

The vignette from my fieldwork is one example of how Bosnians debate displays of religion, that is, the visibility of religion in 'public' spaces and through these debates relay their understanding of what is, or should be, 'appropriate' religious expression consistent with national belonging. The ways in which Mirsada and her family dealt with issues of the visibility of religious symbols revealed their attempts to negotiate the possibilities for inclusion and create fields of belonging for themselves.

Mirsada's expression of religious identification entailed negotiating across and within different fields of belonging. Mirsada had stated to me on previous occasions that she was not a practising Muslim and as such did not feel that the mosque was where she belonged, as it was a place of belonging for pious people. However, on occasions such as the commemoration we attended, Mirsada was able to conceive of this site as a space of belonging, revealing that social spaces are sites of contestation in which boundaries of inclusion and exclusion are continually drawn. In this way, local spaces are contested where the lines of belonging can be articulated differently on various occasions. Thus, establishing fields of belonging, as Gedalof (2003) argues in relation to the creation of 'homes', is never complete but rather is a process of continual production and reproduction. The active and creative ways in which this is experienced is evident in Mirsada's experiences at the mosque. The occasion marked a significant event in Bosnia's history and as such the gathering at the mosque promoted an actively ethnic rather than simply religious belonging with which Mirsada could identify. Thus, throughout the afternoon, Mirsada enthusiastically participated with the other attendees in conversations about shared experiences in Bosnia and Australia. By embracing an ethnic Bosnian Muslim identification, she established fields of belonging within the spaces of the Islamic Centre. It was in this setting that she and I were diligent in ensuring that our heads were covered before entering the mosque, as we were fully aware that this was an expected practice. While the headscarf was not something either of us ordinarily wore, on this occasion Mirsada embraced it as a marker of inclusion. Thus, it was by knowing and adhering to these material practices that Mirsada claimed her place within this field of belonging. However, Mirsada consciously decided not to wear her headscarf outside the confines of the Islamic Centre for fear that in other local contexts it would become a marker of exclusion. She was keenly vigilant not to put her headscarf on at home lest her neighbours see. Thus she actively intended to privatise religion, to keep her religious expression visible only within select spaces. This act conflicted with Mirsada's pronouncements that the freedoms of religious expression were central tenets of Australian secular belonging.

Mirsada's censoring of her religious identification was oriented largely in relation to her neighbours, in particular those she described as 'Australian'. Such sentiment revealed that Mirsada felt her citizenship was not equivalent to that of her neighbours and relayed her understanding of the differential ways in which citizenship and belonging were claimed. Her actions spoke to a sense that she was less entitled to claim belonging, whilst they were the ones who could bestow or deny her inclusion. Thus, it was their approval that Mirsada required in order to claim her own right to belong as a citizen and, in this case, outward Muslim identification was perceived to create obstacles for this form of belonging. Ajrouch and Kusow relay a similar incident occurring to Joumana, a Lebanese Muslim immigrant in the United States of America, whose decision to adopt the hijab resulted in a negative reaction from her neighbour. In that case, the announcement of a religious identity through dress challenged Joumana's position as a full citizen (Ajrouch and Kusow 2007: 85–6).

Mirsada's vocabulary spoke to the different spheres of belonging she maintained even with her neighbours. She referred to the 'migrants', such as the old Greek widow, with whom she openly expressed her ethnic and religious identification, as opposed to the 'Australians', from whom her religion was privatised. In this way, citizenship to a secular multicultural nation was embraced through local assertions of mutual difference which enabled forms of intersection to be identified – this was promoted in her engagement with the 'migrants' in her neighbourhood. It was with them that she identified and hence with whom she was able to stake a collective claim to both local and national inclusion. Conversely, she was more reserved with the 'Australian' neighbours – who she judged to be more entitled to dispute and challenge her claims to belonging. She felt constrained in expressing her Muslim identification with her 'Australian' neighbours and this is conveyed in the active negotiation of her religious expression. Such judgement was based on her prior interactions with these neighbours, who she identified as being very 'nationalistic'. She suggested that it was on account of her 'whiteness' that they had felt comfortable sharing racist thoughts with her about the Vietnamese neighbours. As a result, she felt that negative comments about Muslims were not too distant.

Mirsada's negotiation of religious expression reveals her engagement with contemporary discourses which position Muslims as problematic citizens in Australian society. This reveals that her actions of belonging became 'both a goal to strive for and one to resist' (Rosaldo 2003: 3). In this way, the objectification of Muslims as problematic resulted in Mirsada's attempt to not belong overtly to this Muslim identification and rather seek a sense of belonging along other lines, because being Muslim was a possible affront toward belonging in Australia.

In addition, Mirsada had internalised the hegemonic portrayal of the hijab as a marker of otherness and exclusion, a fact reflected in her attempt to censor her religious expression. Mirsada was not adverse to 'public' expressions of religion, as indicated earlier, in relation to her neighbours' willingness to express their faiths. However, she was conscious of the differential judgments about 'acceptable' forms of religious identification in public spaces, particularly those directed toward Islam.

While Mirsada did not state it, I suspected that her concern in displaying the hijab to her neighbours was not restricted to the 'Australians'. As noted earlier, multicultural belonging in Australia is based on an ethnicised approach to religious identification. This is conveyed through the various religious organisations that are structured along ethnic lines. Thus temples, churches and mosques often carry an ethnic affiliation (Humphrey 2005: 136), as I noted earlier in the description of the religious institutions in Mirsada's locality. This was clearly evident in the mosques established to cater for the Bosnian communities that I observed, where the leadership almost exclusively comprised of Bosnian migrants, and where the practices and even architectural styling were informed by the Bosnian context. It was such ethnic Bosnian Muslim identification that Mirsada felt comfortable embracing within the spaces of the Islamic Centre. Also, it was this ethnic form of religious identification that she shared with her 'ethnic' neighbours. Indeed, in speaking about her neighbours' religions, she would prefix it with their ethnicity, such as Greek-Orthodox.

The hijab, however, in recent years has become a symbol of a post-ethnic religious identification with Islam (Humphrey 2009), and as such in the contemporary politicised discourse, it represents a challenge to belonging in Australia. Such identification challenges the conceptual bounds of ethnic religious identification within the multicultural paradigm. It is this positioning of the hijab as a threat to multicultural belonging, I suggest, that also informed Mirsada's negotiation of its use.

Debating Public/Private

The second incident relayed in the vignette from my fieldwork diary, above, focused on my informants' reactions to the newsreader's cross. The intensity of both Mirsada and Sedin's reactions to seeing the cross on the newsreader's necklace, clearly suggested that they felt confronted by the presence of a religious symbol on a news bulletin. Their reaction was prompted by their expectation that the news should be a reflection of society, one which they understood to be based on secularity. The visibility of the cross disrupted this image and hence contested their expectations of the bounds of the national field of belonging.

My participants claimed belonging in Australia by embracing secular living. They argued that religion should be a 'private thing', and to belong to a multicultural secular Australia meant adhering to these expectations. Their understanding was that religion could be freely practised, without discrimination, but there were limitations. These limitations meant that religion should be kept in private spaces but that keeping religion 'private' meant different things in different settings. Primarily, it meant that religion should not be disruptive to other people's belonging and that religious neutrality should be maintained in schools, in places of employment and, above all, in government institutions and the media. Thus, their objection was to the dominance of, or the acceptance of, only particular religions within the national frame. They contrasted the institutional dominance of Christianity with the everyday acceptance of religious diversity within multicultural spheres.

While Mirsada found the newsreader's religious symbol in a public forum confronting, she was not perturbed by other 'public' expressions of religion that she encountered in her local neighbourhood. Thus, Mirsada's experiences of everyday multiculturalism suggested that not all 'public' displays of religious expression were discouraged, but that instances where this challenged or threatened belonging were indeed problematic.

As indicated earlier, Mirsada's neighbourhood was composed of people with various ethnic and cultural identities. Here, it was migrants who constructed the multicultural experience, and their expressions of their diverse religiosity did not threaten Mirsada's sense of belonging to a secular multicultural society, but instead reinforced it. This everyday multicultural reality allowed and fostered an exchange with people of different religions and ethnicities and this diversity was consistent with her claims to Australian citizenship. These exchanges were not considered problematic because they did not impinge on her sense of belonging, but rather enhanced it, since belonging was based on an acceptance of religious freedom. It was within such settings that Mirsada did not feel constrained in wearing her own religious symbol, a pendant of a crescent moon and star. Thus these 'public' expressions of religion in suburban spaces were not considered inappropriate as they remained consistent with the daily interactions which promoted belonging along multicultural secular lines. However, public expressions of religious identification in media spaces were confronting, because they reaffirmed the dominance of Anglo-Christian[3] identification in Australian institutional spheres, consequently disrupting the national multicultural imaginary.

Mirsada's varying acceptance of 'public' expressions of religious identification reveals that definitions of 'public' and 'private' are open to interpretation and contestation, where 'public' does not always refer to a particular place or sphere of activity. This demonstrates, as Gal (2002) argues, 'public' and 'private' are constituted contextually and can be found within each other's realms, thus private

3 'Anglo' is often used interchangeably with 'Anglo-Celtic', 'Anglo-Saxon', while these categories subsume internal ethnic and religious differentiation, they are commonly used in public discourse to refer to Australians with some British background. My use of these terms reflects the manner in which scholars have sought to identify forms of hegemonic belonging based on ideas of 'whiteness' and 'Anglo-ness' (Stratton 1999).

spaces can be made in public realms and vice-versa. This is clearly seen in the earlier example where Mirsada censored her religious expression, consciously waiting to be in the confines of the Islamic Centre before putting the headscarf on, lest her Australian neighbours see. In this instance, the private home space became a public space from which her Muslim religiosity was to be hidden. However, even within this space, different fields of belonging were enacted with her other neighbours, so with Sofia, the Orthodox Christian Greek widow, Mirsada was comfortable wearing her Islamic pendant. Mirsada's demarcation of such spaces speaks to the migrant critique of their multicultural reality and Australia's secularity.

The local park was another site where 'public' and 'private' forms of religious belonging were differentially constituted. The public park was managed by the local government and was accessible to all residents. It was a common socialising space, particularly on weekends as it offered picnic benches and barbeque areas, a children's playground and walking tracks. On occasion, the park would be reserved for celebrating special events, such as the Hindu festival Holi, Christmas carolling and for the Muslim festival of Eid. Sedin and Mirsada appreciated such 'public' events as examples of the freedom of religious expression that Australian secularism promoted. Their appreciation, however, was based on the understanding that such festivals were in fact located in 'private' spaces. Thus, the public space of the park was reconfigured as a 'private' space for religious identification.

Sedin explained that attending an Eid festival at the local park differed from praying in that same park. In the former scenario, the public arena becomes privatised; as he noted, 'Last month I took my children to the Eid festival. I think that that's fine. Even though it's in public, everyone knows that it's a festival, we're not interrupting others' activities in the park.' He differentiated these displays from seeing a group of men praying in the park. Here his attitude differed: 'When I see people praying in the park, I don't know. I don't like it, I wouldn't do it. I wonder what the Australians who are there, I wonder what they think.' Praying[4] in the park was considered a public activity. The park, in this sense, remained a public arena and the act of praying was seen to conflict with what was appropriate religious expression. Much like Mirsada's reference to her neighbour, Sedin was concerned about what 'Australians' thought; from previous conversations with him, I understood this to refer to persons of Anglo-Celtic heritage. Once again, the negotiation of one's religious identification was made in reference to those who were perceived to be the dominant national group in Australia and who were able to challenge Sedin's claim to belonging. However, in order to achieve a form of belonging, Sedin enacted practices (or rather refrained from them) in order to claim his status as a citizen. His local encounters suggested that it was not simply about

4 Practising Muslims are expected to follow the Islamic ritual prayer five times a day, according to prescribed times. The prayers consist of a series of positions which include standing and bowing movements.

'concealing' his prayers and his identification, but rather a matter of determining when and in which social circumstances such expressions were acceptable.

The 'public' local encounters with religious expression varied from the 'public' expression of the cross on national television. The latter was interpreted as an exclusionary sign, one which challenged the plurality of Australian belonging and worked to reaffirm the sentiment of Christian Anglo-Australian dominance. As Mirsada and her family were Muslim, the newsreader's cross served to challenge their belonging. However, it was not only the presence of a Christian symbol, but also the interpretation of the newsreader as an 'Australian', an 'Anglo', who embodied the Australian nation, which worked to challenge Mirsada's sense of belonging to Australia and relegated her into the migrant/ethnic/outsider space. Indeed, while Australian multicultural policies enable a freer expression of cultural and religious difference, limits and contradictions to multiculturalism remain. While the appreciation of difference, often typified by 'migrants' and 'migrant cultures' is fostered by governmental policies, ethnic bias in terms of Anglo-Australian dominance continues and extends into institutional, legal and political spheres (Hage 1998, Stratton 1998). In this way, while the migrants come to embody the multicultural reality, they remain outside of full national belonging.

These contradictions lead to varying experiences of citizenship which can result in feeling like, or being treated as, a 'second-class citizen' due to one's positioning in relation to race, ethnicity, class and, in this case, religion. So Mirsada's critique of her own belonging in her local multicultural setting is contrasted with the realisation of an Anglo-ethnic and Christian dominance in constructions of full citizenship. Thus, Mirsada and her family members' objections to the newsreader's cross constituted a protest against hegemonic constructions of Christianity as normative in Australian society, which consequently placed other religions outside the 'mainstream' sphere, as revealed in Jasmina's remark, 'It's their country'.

Cognisant of these conflicting messages surrounding the parameters of belonging, Mirsada conducts her own critique of Australia's secularity. In conversations with me, she voiced her frustration that public expressions of Islam had been isolated from other religious symbols, and had been presented as particularly problematic for citizenry belonging. On the day of the fieldwork vignette presented above, she referred to the newsreader's cross in questioning the parameters of belonging in a secular society. She stated: 'Why are they able to wear *that* but they create a fuss about girls wearing the hijab to school? I mean isn't religion meant to be a private thing anyway? Don't we live in a secular society?' Mirsada's questioning reveals her uncertainty that belonging in Australia is truly based on adhering to secular values. Indeed, her frustration at the differential acceptance of public displays of religious identification works to unsettle her claims to belonging on both local and national levels.

Conclusion

Belonging is a dialectical process of migrant engagement with discourses of inclusion. Through this engagement, migrants negotiate their own national and citizenry belonging. The discussion in this paper has demonstrated that in claiming citizenship, one seeks recognition as a rightful member of a group and such desires for belonging are enacted within everyday environments. Thus, migrants engage with debates and discourses which contest the meaning and parameters of inclusion within various social spaces. Such engagement reveals the ways in which migrants become emplaced in the national and local, as it is through debating, challenging and negotiating the bounds of belonging that they seek to establish spaces for themselves. In this way, sites such as the local neighbourhood, local streets, parks and religious centres are the arenas through which the parameters of national, ethnic and religious belonging are negotiated.

For Bosnian Muslim migrants, claiming belonging within the Australian multicultural secular framework has entailed oscillating between various forms of 'public' and 'private' displays of religious identification. Such negotiations are informed by various spheres of Australian society which convey divergent expectations about religious expression and its role in sustaining fields of belonging. Bosnian Muslims have predominately sought to claim belonging by embracing ethnicised religious identification where public displays of religion are enacted as part of a cultural identity. In such cases, being 'Muslim' is prefixed by being 'Bosnian'. In this way, Bosnian Muslims are able to claim full belonging as citizens according to Australia's secular multicultural paradigm. This project of asserting cultural-religious identification, in order to stake a claim within Australia's multicultural imaginary, has been further propelled by the recent public and political rhetoric which has questioned the position of Muslims within the national secular space. This is because being 'Muslim' without an ethnic identifier has been configured, in public and political debates, as an affront to Australia's secular framework. This objectification of Islam and Muslims as problematic to belonging has heightened Bosnian migrants' scepticism about the status of Australia's secularity and brought into question the place of religion in framing the bounds of national inclusion. Indeed, this has prompted their own critique and judgement about the ways in which secular multicultural belonging is and could be framed. Such active reflection, negotiation and critique reveals the ways in which migrants become emplaced and come to utilise various encounters, such as those within their local community, with media representations and political discourses in order to determine and challenge their positioning within various fields of belonging. In this way, the bounds and expectations of secular multicultural living are revealed to be the subject of continuous negotiation.

References

Ajrouch, K. and Kusow, A. 2007. Racial and religious contexts: situational identities among Lebanese and Somali Muslim immigrants. *Ethnic and Racial Studies*, 30(1), 72–94.

Castles, S. and Miller, M. 2003. *The Age of Migration.* New York: Palgrave Macmillan.

Gal, S. 2002. A semiotics of the public/private distinction. *Differences: A Journal of Feminist Cultural Studies*, 13(1), 77–95.

Gedalof, I. 2003. Taking (a) place: female embodiment and the re-grounding of community, in *Uprootings/Regroundings: Questions of Home and Migration*, edited by S. Ahmed et al. Oxford and New York: Berg, 91–114.

Gupta, A. and Ferguson, J. 1997. Discipline and practice: "the field" as site, method, and location, in anthropology, in *Anthropological Locations: Boundaries and Grounds of a Field Science*, edited by A. Gupta and J. Ferguson. Berkeley: University of California Press, 1–46.

Hage, G. 1998. *White Nation.* Sydney: Pluto Press.

Humphrey, M. 2005. Australian Islam, the new global terrorism and the limits of citizenship, in *Islam and the West: Reflections from Australia*, edited by S. Akbarzadeh and S. Yasmeen. Sydney: UNSW Press, 132–48.

Humphrey, M. 2009. *The Domestication and Securitisation of Muslims and Islam as a National Security Project.* Paper to the International Conference of Migration, Citizenship and Intercultural Relations, Deakin University, Melbourne, 19–20 November 2009.

Lovell, N. 1998. Introduction, in *Locality and Belonging*, edited by N. Lovell. London and New York: Routledge, 1–22.

McAuliffe, C. 2007. Popular cultural relativism: religion, racial profiling and national multiculturalisms, in *Everyday Multiculturalism Conference Proceedings*, edited by S. Velayutham and A. Wise. Macquarie University: Centre for Research on Social Inclusion.

Modood, T. 2007. *Multiculturalism*. Cambridge: Polity Press.

Olwig, K. 2002. The ethnographic field revisited: towards a study of common and not so common fields of belonging, in *Realizing Community*, edited by V. Amit. London and New York: Routledge, 124–45.

Olwig, K. and Hastrup, K. (eds) 1997. *Siting Culture: The Shifting Anthropological Object.* London and New York: Routledge.

Ong, A. 1999. *Flexible Citizenship: The Cultural Logics of Transnationality.* Durham and London: Duke University Press.

Ong, A. 2003. *Buddha is Hiding.* Berkeley: University of California Press.

Poynting, S. and Noble, G. 2004. *Living with Racism: The Experience and Reporting by Arab and Muslim Australians of Discrimination, Abuse and Violence Since 11 September 2001*. Report to the Human Rights and Equal Opportunity Commission, Australia.

Rosaldo, R. 2003. Introduction: the borders of belonging, in *Cultural Citizenship in Island Southeast Asia*, edited by R. Rosaldo. Berkley: University of California Press, 1–15.

Stratton, J. 1996. The impossible ethnic: Jews and multiculturalism in Australia. *Diaspora*, 5(3), 339–73.

Stratton, J. 1998. *Race Daze*. Annandale, NSW: Pluto Press.

Stratton, J. 1999. Multiculturalism and the whitening machine, or how Australians become white, in *The Future of Australian Multiculturalism*, edited by G. Hage and R. Couch. Sydney: The University of Sydney, 163–88.

Triandafyllidou, A., Modood, T. and Zapata-Barrero, R. 2006. European challenges to multicultural citizenship, in *Multiculturalism, Muslims and Citizenship*, edited by T. Modood et al. London and New York: Routledge, 1–22.

Yuval-Davis, N., Anthias, F. and Kofman, E. 2005. Secure borders and safe havens and the gendered politics of belonging: beyond social cohesion. *Ethnic and Racial Studies*, 28(3), 513–35.

Chapter 10
Migrants on Campus: Becoming a Local Foreign Academic

Brigitte Bönisch-Brednich

The first term that I taught at Victoria University of Wellington, New Zealand, was full of surprises and misunderstandings; some were followed by dismay, some by joy and others I only identified long after the event, when everybody else had forgotten my incomprehension. Many of my migrant colleagues now tell me they too were caught unaware by how much there was to learn, to think about and, above all, how different the expectations of pedagogy and scholarship were. I wonder now about the enthusiasm and the innocence that guided and guarded me back then. It helped me through the challenges of becoming a senior staff member while having no idea what a course outline was or why I should let others know what my course readings would be, and the puzzlement and frustration I felt when I realised I was not free to decide what to teach and was expected to apply to a whole raft of committees if I wanted to make curriculum changes for the following year. The binding pre-course agreements on assignments, readings, tutorials and lecture outlines seemed an intrusion into my personal sphere as a university teacher that I found hard to understand.

There were many more puzzles: Why did people work with their doors open? Why did students never come during my office hour, but instead just walked through my door whenever it suited them? Why were staff meetings conducted in a stiflingly formal atmosphere if everybody was supposedly on an even footing? This was particularly surprising; after all, for me one of the main attractions of New Zealand had been the perceived lack of the strict hierarchy that characterises the German university system. And how did problems get solved, if there was always an empty silence when I voiced them?

The campus, at first sight a familiar territory, remained for a long time a physical and social maze that only revealed itself step by step after joining working groups and committees and attending union meetings (unions being something unheard of where I came from). It also lacked the social centre I was used to, the campus hub with huge eating halls where three-course lunches were served to thousands of students, where espressos and gossip were consumed, books were bought and doctor's appointments made. To me it was a wonder that students did not starve, if all that was available was a few sandwiches.

Other problems were more personal: I had always assumed that my English would be good enough, but I was stunned by how difficult it proved when teaching huge first-year courses. I wondered whether my voice would ever be clear enough, how I could enthuse students, make them laugh, shock them playfully and be stern when necessary, when all I could focus on was somehow getting my content across. Other migrant colleagues were incensed that nobody would pronounce their names correctly. 'They do not even try', they said, although this did not bother me greatly; rather, I pitied everyone who looked at my name and despaired. What did bother me was the slow realisation that the publications of which I was so proud seemed to count for little, that I was now in an environment where hardly anyone could read my work and that I was living in an evolving auditing system that did not know how to judge it. There were other challenges: learning the rules of self-management and how to express myself when negotiating the university bureaucracy and reporting systems. Being a woman, there were the added complications of how to deal with the strange, recurring misogynistic tactics I encountered and had no armour against, and often only detected long after the event.

Sometimes the everyday social workings of the campus felt like a series of castles with bridges drawn up; sometimes you would discover new facilities, such as the staff club – an endangered but still existing locale of New Zealand academic social life. Often, my anxiety was accompanied by the joy of having a great job and, even more, of being a cultural anthropologist. This is fun, I would think, this is exciting, let's decipher this, let's rehearse new-found rules of conversation, let's try negotiating something with a newly grasped social technique. I called such trial-and-error sessions my auditions for personal acceptance.

The Strangers on the New Zealand Campus

Fortunately for me, New Zealand universities are full of strangers on campus. With the growth and expansion of the global knowledge economy, international mobility of students and academics has become an expected and encouraged pattern in academia. The percentage of international faculty is one of the quality factors measured when ranking universities. New Zealand has one of the highest numbers of migrant scholars in the world, with up to 63 per cent of staff being international academics, yet no research has been done as to how this came about and, more relevant to this text, how these migrants arrive and integrate on a New Zealand campus.[1]

New Zealand universities have always depended on overseas hiring, the first scholars coming from Scotland and other parts of Britain. Unravelling the complex

1 Research has been done on academic migrants in other countries but, to my knowledge, has mainly been conducted in the areas of commerce and higher education. Consequently, the focus has been on career analysis and problem solving in relation to pedagogical issues arising for migrant scholars (see Richardson and McKenna 2002, Scheibelhofer 2003, Kolapo 2009).

issues of New Zealand's historical Commonwealth ties and discussing the impact of neoliberal hiring strategies on recruiting excellence is too complex to address fully here.[2] I will, however, try to discuss what it means and feels like to arrive and work in such a multinational workplace, which is nevertheless still located and deeply rooted in the New Zealand university system, which is engaged in nation building and the development of a local knowledge economy.

Since nearly half of the academic staff at Victoria University of Wellington are from outside New Zealand, first impressions are puzzling. The university is organised so as to accommodate international networks, despite being embedded in the New Zealand education system with a commitment to local students and a history of strong ties to British academia. Many departments and schools also have close connections to universities in Australia and the United States; many of the New Zealand-born scholars have returned after finishing their doctorates from Oxbridge, the Ivy Leagues or 'across the ditch' in Australia. Looking at the many different gowns being worn at the first graduation ceremony I attended in New Zealand – the first time in my life that I had worn a gown and attended a graduation ceremony – was an introduction to the multiplicity of scholarly backgrounds, international networks and academic attitudes to scholarship. The American gowns seemed baroque, opulent and in-your-face; the Oxbridge ones restrained, but worn with a habitus of pride and belonging; the Australian ones light and playful versions of the American ones. My own gown comes from a southern German tailoring business, which reproduced a model borrowed from the archives of my alma mater. Students saw gowns as representing paralysing hierarchies and they were abolished in Germany in 1968. Donning one, learning to wear it and accepting it as part of the university culture seemed to encapsulate my academic migrant experience. The varied gowns of my new colleagues mirrored the differences in their accents and dialects when they spoke English. Some, like myself, were clearly from a non-English background, but while others sounded as if English was their mother tongue, my untrained ears had difficulty distinguishing between, and sometimes even understanding, them.

Mastering the art of detecting accents provided a way of feeling *in* place rather than standing aside and observing myself and others, like the enduring stranger in Alfred Schütz's celebrated essay on what it means to be a migrant. Schütz's analysis of what it takes to learn to be the foreigner, the 'Other', is enlightening because it allows for both the growing understanding of the new society and the ongoing need to negotiate one's life inside and outside that new culture: '[the migrant] has to question what seems unquestionable to the in-group and cannot put his trust in vaguely knowing about the general style of the pattern but needs explicit knowledge of its elements. This entails a dislocation of the stranger's habitual system of relevance. A thorough modification of his schemes of orientation

2 For a discussion of some of these issues refer to Bönisch-Brednich (2011, forthcoming).

and interpretation and his concepts of anonymity, typicality, and chance is the prerequisite for any possible adjustment' (Schütz 1944: 499).

Having learnt to be 'the Other' means that I can never be Kiwi, nor do I aspire to be; after some years, however, I and other academic migrants like me may even become incapable of re-immersing ourselves in the academic worlds from which we came. We become the Other in both worlds, and as a result will always be reflexive – though not always consciously so – about our place in academic environments. Never again will I have the knowledge that I just belong, a sense that came from subconsciously acquiring the social rules of distinction first as a student, then as a locally socialised academic being on my home campus. We migrant academics find that the finer points of academic life, so necessary for survival and success, must be relearned, often painfully. We cannot identify the academic aristocracy of this new country, nor can we immediately grasp the tactics of class or social belonging, or identify if and how class conflict, racism and gender politics are played out. We enter a social vacuum where mastering language is the foremost task, and academic achievements are at the centre of our efforts. Having experienced this, we will never be able to accept these aspects of academic life as a given, seeing them simply as 'the way things are done around here'.

The Second Birth

With this short, partly auto-ethnographic, tour of the campus as a locale where migration is lived and negotiated, I have attempted to focus my discussion and anchor my exploration to guide both myself and the reader in considering individual and collective experiences on a multinational campus, as well as wider migration issues and the global development of academic workplaces.

This chapter is based on informal conversations with 15 academic migrants, mostly working within the New Zealand context, and on field notes from my own migrant experience and my history of researching migrant topics. The analysis is, therefore, informed by my observations and interviews while trying to unravel the cultural mysteries of New Zealand as a place for settlement.[3] I write as a migrant with English as my second language, and very much from a humanities and social science perspective.[4] Being German and migrating from a continental European pre-Bologna university system into one formed on the basis of the

3 I see my personal text very much embedded in information from other academic migrants' experiences and the narratives they have offered me over the past year. A variety of publications on auto-ethnography have been very inspiring in conceptualising my ongoing fieldwork and the writing of this chapter (see Anderson 2006, Ellis 2009).

4 This project is a continuing one, and I am looking forward to interviewing academics from other disciplinary areas, such as the sciences, in order to examine the ways in which their experiences might differ from mine and those of my more immediate colleagues. I believe it will also be fruitful to pay special attention to English-speaking academic

British understanding of tertiary education, I experienced a much greater cultural shift than I had imagined I would. I was doubly unprepared for the surprises and difficulties I encountered when I settled long term. Despite my one-year residence in New Zealand when I carried out a major research project before moving here permanently, I found having been hosted and treated as a passing visiting academic had allowed me the comfort of being able to ignore the intricacies and duties of daily life, and did not prepare me for the challenges of being involved in teaching and administration. In retrospect, much of my research has been engaged with ethnographically examining the processes of settling and fitting in.

Making a home on campus is a central part of everyday academic life. Although much of our disciplinary identity might well be bound up in international networks, most of our working lives are spent within the spatially bounded sites of a university campus. In his essay on place-making as a project, John Friedmann describes the university as one of the many work places which are 'structured by relations of power among individuals and institutions that occupy strategic positions within it'; therefore 'the accumulation of 'symbolic capital' ... corresponding to each field [inside the university] is a characteristic habitus which gives each player a 'pre-reflective, infra-conscious mastery' of the field' (Friedmann 2005: 316). Although this is just a passing remark in his discussion of multiethnic urban centres, he describes the university campus as a place where locality is many things at once: a 'real', physical place of residence and a location where symbolic, political and cultural capital is a basic necessity to maintain and improve social positioning. Using Bourdieu's notion of habitus, Friedmann discusses the need to allow for more dynamic change and adjustment in the paradigm of social disposition. 'It happens to most of us, that we are forced (or may choose) to enter another field than the one in which we are 'at home'. This involves learning a new set of rules, a process that is slow and painful and which we experience, Bourdieu asserts, as a kind of 'second birth' (Bourdieu 1990: 58 in Friedmann 2005: 317).

Although Bourdieu's mention of this 'second birth' is made within the context of his discussion on processes of shifting class, it remains a helpful way of reflecting on migration. When thinking about academic migration, it is also useful to refer back to such well-known theories and tables of the migration process as immigration, acculturation and adaptation, or as periods of honeymoon, marginalisation, segregation and (re)integration (Berry 1997, Kreutzer and Roth 2006). However, we quickly realise that, although there might be good templates for theorising the course of migration and the change of social positioning, these remain generalisations and do not address if and how specific migrants or groups of migrants move through such phases. A focus on the subjective and specific local experience can enrich and alter such sweeping academic metaphors of change.[5]

migrants who do not change language while migrating but nevertheless have had to cope with an even more pronounced expectation of 'similarity'.

5 For a discussion of the connection of 'making home' and movement and migration see Rapport 1995 and Rapport and Dawson 1998.

Assumptions of Similarity and Experiences of Difference

Before we can come to understand another culture, we must first experience it as an 'Other', something strange and perplexing (Bredella 1991: 23). In an essay on 'defogging expatriate expectations', Teresa Ting recounts her experiences in a south Italian university, where local networks over-ruled academic excellence. Only after some painful years and shifts in her career did she come to understand the different yet immensely powerful cultural logic of the Italian academic system. In order to understand why they were not welcomed with open arms, she and her co-author learned 'to negotiate the different matrices of power we encountered … [the] need to consider the nature of the communities we were trying to enter and that defined us as strangers, as well as the capital we carried to those communities' (Ting and Watts 2009: 93).

Every migrant academic I spoke with has similarly insightful anecdotes and observations on their experiences. One female scholar in senior management, who had just recently arrived from Australia, said she had found it tempting to imagine New Zealand universities to be very much the same as those in Australia, but soon discovered that they are actually quite different (personal communication, March 2009). Rather than entering immediately into a discussion on what these differences are, it is more rewarding to begin by exploring the first part of her statement: 'it is tempting to imagine it to be similar'.

This encapsulates one of the striking findings of my investigation so far: the nearly axiomatic assumption that universities – at least the ones in the developed world – are very similar. This seems surprising, because migration is usually carried out with an expectation of difference and some difficulty; following the so-called 'honeymoon phase' after arrival, this difference becomes something to resent, battle with and attempt to resolve. So why would scholars migrating to a campus in a new country not expect things to be different? This imagination of similarity sets academic migrants apart from other migrant groups, and is an assumption that seems to be shared by both the academics and their universities.

It is understandably tempting to make this assumption of similarity. The basic structures of the overall organisation and the academic workplace seem similar: we have faculties providing a roof for bundles of disciplinary orientations; we have the obligation to teach, research and administer; there are internal and external funding bodies which are linked to a common and well-known pressure to publish widely and teach wisely; and there is the commonly acknowledged global change to the essential character of universities imposed by governments following neoliberal ideologies. Most academics will also assume that, wherever they move, they will be at home in their own discipline with colleagues who share a disciplinary background, and they always hope that any differences from their old university will be positive ones. Migrants always hope that the academic lawns are greener on the other side; after all, that is why they migrate, and they certainly expect New Zealand to be greener, cleaner and wonderful. They often focus their worries on how their families and spouses might cope in the new environment and with possible language barriers outside the perceived safety of the university campus.

Such guiding assumptions of sameness, however, hinder academics' ability to reflect on the meaning of migration and the requirements of successful migration within the university setting. Migration always means uprooting, culture change and culture shock. Bringing migration into the context of the cultural and social procedures of making home, of becoming local, sets up a task requiring an analysis of human movement that is more often than not unexamined and uneasily narrated. Theodor Adorno, one of the most famous academic exiles, reflects in his *Minima Moralia* on his migrant experience in the United States and declares that 'every intellectual in emigration is, without exception, mutilated, and does well to acknowledge it to himself. He lives in an environment that must remain incomprehensible to him ... he is always astray ... His language has been expropriated, and he is cut off from the historical dimension that informed his knowledge' (Adorno 1951, corrected translation).[6]

The academic migrants I talked to reported this feeling of being lost easing with time, but migrants seldom explore the deeper meaning of the cultural logic of their new country, they tend to just become enculturated.[7] Often there is lasting incomprehension as cultural problems are perceived as personal problems, and the new workplace is constantly compared with the old workplace, often without reflection on the underlying clash in 'cultural grammar' (Giordano 1996: 34ff). This, then, often results in either return migration or isolation and social dysfunction. Most departments in my faculty have a history full of promising academic appointments, where the new appointee then departs after two or three years leaving behind students, a lasting sense of bad relations with management and an unspoken but seething disappointment. Other departments are stuck with colleagues who arrived with high expectations, but stayed despite never really accepting or understanding the local workings of the Kiwi campus, resulting in bitterness and developing poisonously altered personalities. It therefore seems worthwhile to reflect on the ways in which the initial years of learning to be local are experienced.

Beyond these core issues, I observe a tendency among my fellow academic migrants to return to some assumptions we had to painfully unlearn in our initial 'second birth': the initiation on our home campuses into our own national disciplinary contexts. When shifting from a student or a post-doctoral position

6 'Jeder Intellektuelle in der Emigration, ohne alle Ausnahme, ist beschädigt und tut gut daran, es selber zu erkennen, wenn er nicht hinter den dicht geschlossenen Türen seiner Selbstachtung grausam darüber belehrt werden will. Er lebt in einer Umwelt, die ihm unverständlich bleiben muss ... immerzu ist er in der Irre ... Enteignet ist seine Sprache und abgegraben die geschichtliche Dimension, aus der seine Erkenntnis die Kräfte zog' (Adorno 1951: 32f). I am grateful to Patricia Nickels for locating this reference.

7 I have to thank Simon Bidwell for developing this point with me in one of our long discussions in my 2009 honours class.

to being a professional academic, young scholars are compelled to come to a realisation of the university as a workplace rather than a place of abstract, non-vocational learning. As students we are raised to believe that scholarly success is based on academic merit, and that excellence is what counts and is needed for an academic career. This concept of meritocracy is contrasted with the realities of environments often characterised by nepotism or an almost feudal domination. In a slow but necessary defogging process, young professional academics acquire more skills and a better grasp on the reality of academic life. We learn that, yes, academic achievements are vital, but equally essential are the abilities to build local and global networks, to speak publicly, to teach, to develop administrative skills, to learn how to apply for grants and to gain basic skills in promoting ourselves and our students. In short, by living through this phase of disillusionment we learn how to function on campus as a professional academic rather than an isolated doctoral student. In Bourdieuan terms, we develop a 'pre-reflective, infra-conscious mastery' of life on campus. When changing campuses and countries, all these professional skills need to be relearned and re-acquired in localised campus contexts, even if this is not initially realised as a priority. The acquisition of language and the immediate and pressing need to learn the general structure of teaching and administrative duties push such relearning onto the back burner, where it tends to stay for much longer than is good for us. Often, we do not even recognise this as a problem, as a long list of other new priorities obscures the lack of these skills as an obstacle to our successful settlement.

Often, in order to cope with the overwhelming tide of demands, migrants tend to ignore or simplify a raft of unexplained observations. One tactic is to resent the unfamiliar local rules necessary to settle successfully into the new academic environment. Misunderstandings or the lack of cultural translation skills can be seen as 'intercultural dissonances caused by diverging "cultural grammars"' (Giordano 1996: 34ff).[8] Forgetting all we have learnt about the development of academic career skills and suffering from the absence of our own networks, we tend to fall back on our student days and revive the concept of pure meritocracy. The new country, then, is often seen as 'corrupt', 'inbred' and 'unfair', with rising resentment of the kind described in migration theory as routinely following the arrival phase. Such feelings of resentment often co-exist with positive imaginings about the new place of residence. For example, academic women often seem to develop a new, positive blindness to glass ceilings, conveniently suppressing the knowledge that male brotherhoods are still immensely powerful and in the majority. This is a product of the period identified as the 'honeymoon', one that tends to end rather more quickly than the subsequent period of 'resentment'. Recoiling into an unreflexive migrant posture can be seen as an understandable resistance to exploring, detecting and reading a new cultural code and comparing it to the well-known rules of the home campus, a knee-jerk reaction to the complex task

8 The original quote is in German, translated into English in Bönisch-Brednich 2002: 176.

of settling anew. Problems and misunderstandings also arise because the cultural significance of specific communication is often inaccessible. These issues are not unique to academic migrants; when interviewing other German migrants, many echoed similar sentiments to one observation that 'when you meet a kiwi, they say *how do you do, how are you going*, but they really do not want to know at all, you are just supposed to say *fine, I am fine*; I find that very superficial' (Bönisch-Brednich 2002: 176f). This complaint was made despite the existence of a similar German greeting, '*Hallo, wie geht's?*', which requires and expects the simple answer '*gut!*' This illustrates a common inability or refusal to translate cultural knowledge; these translation problems, if not deciphered and re-learned, prolong the integration experience.

These issues all hinder what I define as having to learn how to become a successful 'Other'. As academics on a multinational and multiethnic campus, and even more so in circumstances such as Ting and Watt's mono-ethnic one in southern Italy, we have to become an integrated 'Other'. We have to learn how to become both a scholar who knows how to function in that new local campus and in the new national language and, simultaneously, how to keep functioning at an international level on disciplinary platforms that are often truly global. We have to become our own intercultural agents in the best sense: researching the new culture and promoting ourselves in that ongoing venture. This is vital as, on campus, 'cultural illiteracy … prevents the development of more interesting exchanges toward the cultivation of worthwhile cross-cultural relations' (Ting and Watts 2009: 75).

Learning the Language

At the beginning of this chapter I identified language as one of the key issues in migration. Many of my migrant colleagues in New Zealand had stories similar to my own: they expected their English would be good enough, counting on previous experiences presenting in English at international conferences, and they assumed that their disciplinary expertise would ensure that lecturing and publishing would not be too great a challenge. But the difficulties of working in an all-English environment, coupled with the cultural illiteracy discussed above, inevitably leads to problems.

I have heard accounts of suddenly feeling, at least initially, very young and inadequate. However good our English might have sounded in our home country, it proves inadequate for sophisticated academic teaching and discussion. Most of my migrant colleagues strongly feel that they have reverted back to the restricted state of a teenager, while trying to negotiate a highly competitive and demanding work environment.[9] This will ease with time, but might never end completely.

9 I have heard several stories of colleagues' anxieties about speaking in faculty or academic board meetings and of feeling ignored when they talk. When I had a bicycle

There are always moments when we encounter these challenges again, when tiredness or stress result in a thicker than usual accent or we experience difficulties in maintaining our acquired fluency in a second language. It is necessary to accept this language problem as part of the ongoing process of the 'second birth' migration requires. Carolyn Ellis has explored the difficult topic of bodily stigmas, unravelling issues surrounding living with a minor speech impediment while maintaining a normal, even successful, academic career (Ellis 1998). In many ways, living with an accent and slightly restricted English linguistic abilities requires tactics similar to those employed when dealing with such a personal stigma. We have to learn to live with it, to try and improve our skills, but also learn how to react when it is brought up. Many migrant scholars face criticism from students claiming they cannot understand them; this can be painful and is often narrated as part of the initial frustrations of life on the new campus.[10] Also problematic is how New Zealanders often do not correct foreign academics, feeling that polite and collegial behaviour does not allow for intervention on such a personal level, which can then lead to ingrained grammatical mistakes.

Many academic migrants experience the second-language issue as equally demanding when it comes to academic writing. When publishing in English, we generally need academic friends to help us out, to correct and edit our texts. This means we also have to learn to accept a new and often lasting feeling of imperfection and dependence. Although my pre-arrival bibliography was something to feel proud of, most of my books and articles are not in English and hardly any of my new colleagues can read them or would make the attempt. Suddenly, I felt stripped naked of some of my academic regalia, as if I had published hardly anything, and I realised that the academic capital materialised in my offprints had disappeared. It was tempting to fall back on criticising and rejecting my new monolingual workplace. Though exhausting, time-consuming and more difficult with a 'minor linguistic impediment', the task of writing professionally in English had to be mastered; the language of publishing had to shift, at least partially, to that new language and its modes of scholarship.[11]

---

accident and could not type, the university staff nurse suggested my school should organise a voice recognition programme to type for me. The discussion shifted immediately from my injury to the problems my accent may create for such a computer programme. In minutes, my impediment was no longer my physical injury, but rather my perceived thick accent.

10 For a more in-depth exploration of this topic see Lippi-Green (1997).

11 Even in one's own discipline there is the need to adjust to varying modes of writing and presenting in different countries. Americans tend to have a threefold structure of first announcing what they are going talk about, then talking about it, and finally summarising their main points. Germans tend to raise a problem and identify a gap in research; then they will try and discuss as many points as possible, assuming the audience will follow, and, rather than summarising, they will most likely end a paper with a challenge. Equally, modes of conference presentation differ widely; at the 2009 joint conference of English, Australian and New Zealand Associations for Social and Cultural Anthropology, it was noted by many of my local colleagues, that 'the English read their papers, often without

This lack of multilingual scholarship also applies when teaching: to adjust to a monolingual environment, I had to store away most of my preferred and tested readings and find English equivalents. It is important to note that, particularly for scholars in the humanities and social sciences such as myself, migrating may also mean a shift in academic focus in order to engage academically with the new country of residence. This may mean, at the very least, finding new ways of establishing research priorities (for instance, becoming more theoretical rather than working empirically), or researching in new directions that explore the culture, history and social problems of their new country of residence, teaming up with local colleagues, or engaging in comparative projects. In a sense, this is also a process of re-birth, exploring new avenues for research and often initially encountering embarrassments, mildly upsetting comments and setbacks, until we are more familiar with our new foci.

Colloquialisms, speech acts and modes of address in a professional setting also prove complex for many academic migrants. Having come from a country and culture with more formal norms of addressing people, I was initially disorientated by the New Zealanders' love of using first names. In many cultures, formal address indicates and demarcates distance and therefore formalises the personal space between people, whereas first names are seen as indicators of personal and emotional closeness. In Germany, where last names and academic titles such as 'Dr' or 'Professor' are commonly used, I would certainly never address students or colleagues with their first names. In New Zealand, however, first names are almost always used and these linguistic boundaries are instead transferred into more subtle and often puzzling rules of distance. To my foreigner's eyes, this denied social difference and signalled an egalitarian power relationship, but I came to realise that this was counterbalanced by body language, politeness and more subtle ways of showing respect. These subtleties have to be learned and, once mastered, most academic migrants enjoy the easier interaction with others.

This first-name culture can be equally confusing for academic migrants who count English as their first language. An English migrant colleague told me that her most frequently spoken sentence during her first year in New Zealand was, 'I have no idea who you are talking about'. Based on New Zealand's colonial history and its links to British academia, she assumed a great deal of sameness on campus and did not foresee a quite different local academic culture. She noted the very puzzling New Zealand habit of referring to everyone by their first names, including individuals not physically present or, indeed, not necessarily even mutual acquaintances. This resonates with my own experiences as a visiting academic: I was constantly mystified by the incongruity between the high academic standard of discussion and the constant (to my ears nearly flippant) throwing around of

---

PowerPoint', instead relying on polished prose and well-phrased arguments. Australians and Kiwis tended to 'PowerPoint' and speak to their slides (field notes Auckland, December 2009). And as you might have noticed by now, Germans love footnotes to create a running intellectual subtext/commentary to their main text.

first names. It took me a long time to figure out that 'Keith', 'Jamie', 'Jock', 'Claudia' and 'Michael' were not other residents at the Stout Research Centre, but academics who had written the basic and standard works of New Zealand history. Eventually I realised, with a deep sense of relief and pride, that I actually had read most of the books to which my new colleagues referred, except that I identified them by the authors' last names. This conversational habit creates an insider/outsider distinction that induces an intense sense of embarrassment and disorientation. Most Kiwis are not even aware that this is problematic for their migrant colleagues; for them, naturally enough, this is just the way things are done here in New Zealand.

Negotiating the Neoliberal University Environment

To some degree, academics have always tended to migrate, and many of the problems I have analysed here show up in the autobiographical writings of scholars throughout history. There are, however, special demands and challenges to such migration in the current global knowledge economy. There is a widespread shift towards restructuring universities on neoliberal business principles, and many universities are now run along line-management systems, with a CEO-like head appointed by a senate or council, mostly comprising members external to academia.[12] Academics are increasingly required to align their scholarly activities of research and teaching along business modes of planning, acting and reporting. The usual processes of experiencing honeymoon or resentment phases can turn into nightmares of misunderstandings and colliding worldviews on what scholarship entails. Jill Blackmore and Judyth Sachs have analysed the radical shift toward 'performativity' in universities, where being seen to perform is as important as performance itself. They see a process at work that exploits academics' dedication to their profession while at the same time alienating them from their university as a workplace (Blackmore and Sachs 2001). An audit system has emerged, particularly within the English-speaking knowledge economy, which controls,[13] reprimands and ranks academic achievement in ways that are heavily criticised and resisted by older academics but are in the process of being internalised by younger ones as part of the process of entering professional life. Such pressures to become a 'self-managing academic' who adheres to business thinking by modelling activity on the external rules of milestone production are difficult enough for locals who have been slowly coerced into the new regime. For academic migrants, this often creates another layer of culture shock, experienced as a deep intrusion into the academic personality they had developed and, moreover, the imposition of another learning process: a third birth, a re-making of the academic self into a participant

12 For an introduction into doing research on the corporate world, refer to Cefkin (2009), and on restructuring universities Wright and Ørberg (2008).

13 Refer to Shore and Wright (1999), Strathern (2000) and Sparkes (2007).

in an entrepreneurial system of producing and selling knowledge. Resisting this often means a slow or sudden professional death.

This shift to neoliberal systems has particular implications for academic migrants not accustomed to these ideals, particularly when these management systems combine with other cultural differences. One of the essential skills in any professional environment is knowing how to deal with dissonance or conflict and, particularly in the academic environment, how to gain personal or departmental resources. Learning new rules of combat and negotiation are hard in any new country, and New Zealand is no exception. This still being a relatively small-scale society, a rule that often constitutes an obstacle to any real negotiation is that of 'saving face', which is generally more important than dealing with the problem. In a country with such a small population and, consequently, a very interconnected university education system with just eight locations, every academic could be connected to at least 50 per cent of everybody else's local New Zealand networks.[14] This, coupled with a deeply ingrained belief in the Kiwi egalitarian spirit, requires a subtle style of negotiation which is very hard for newcomers to grasp. I have discussed such issues elsewhere (Bönisch-Brednich 2008: 9), but here I would like to emphasise again that it is of the outmost importance to acquire the skills of diplomatic academic conduct. Learning the rules of social interaction the Kiwi way is especially important as, although the university may be populated by international scholars, many university managers are locally born and bred.

It is also vital to acknowledge some of the other underlying challenges inherent in gaining acceptance in a new university environment. For instance, many female academic migrants slowly develop an awareness of gendered career obstacles and eventually learn to detect and deal with the hidden barriers of local and cultural misogyny and our own feminised behaviour. This can mean learning to speak up in meetings, acquiring a cultural and linguistic grammar and literacy for effective female assertiveness, developing humorous ways of dealing with setbacks and conservatism and forming female support cohorts. Most important of all, we have to stop hiding behind our desks, working silently away and dreaming of being discovered and valued by some fantasy academic committee rather than learning some skills of self-promotion.

Negotiating the complexities of a New Zealand neoliberal university environment has been an ongoing challenge for me. When submitting documentation for the audit and ranking systems which now monitor academic outputs, I found myself having to explain my academic life of the past six years into three different categories and several explanation boxes. I asked an experienced friend and colleague for proof-reading and advice. The feedback was curiously unsettling: 'The problem', my friend explained, 'might not be so much

14 In their article on migrant academics at a Sydney university, Saha and Atkinson have argued that 'migrant academics are insiders with respect to the international professional community, but outsiders with respect to the local institutional and organizational community' (1978: 216).

the foreign language publications – those can be hidden away – the problem is that you still have no idea how to write about yourself'. This was referring to my problems in phrasing my responses in such a way as to be appropriately impressive while still being understated and humble enough to suit the taste of the Kiwi judging panel. My friend concluded, 'You simply do not know how to play this game'.

Conclusion

Depending on the reader, all this may sound either extremely familiar and distressing or very alien. However, we need to be able to picture and understand what adjusting to a new system does to scholars, and what lessons there are to be learned here for migration studies. It is necessary, therefore, not to disengage qualitative inquiry and auto-ethnographic writing from the political and economic realities of life on campus. Because academic migrants are compelled to play an important role in that new environment – for migration itself is promoted as a path to a modern university career – it would be negligent not to pay very close attention to how shaky, changing and political the ground on which our campuses are built truly is. Merging research on locality and locale and using qualitative and narrative methods in an analysis of modern university life will enable us to comprehend the tension between local life on campus and the forces of such life being firmly fitted into a global knowledge economy.

References

Adorno, T. 1951. *Minima Moralia: Reflexionen aus einem beschädigten Leben.* Frankfurt a.M.: Suhrkamp Verlag.

Anderson, L. 2006. Analytic autoethnography. *Journal of Contemporary Ethnography*, 35, 373–95.

Atkinson, P. 2006. Rescuing autoethnography. *Journal of Contemporary Ethnography*, 35, 400–4.

Berry, J. 1997. Immigration, acculturation and adaptation. *Applied Psychology: An International Review*, 46(1), 5–34.

Blackmore, J. 1996. Doing emotional labor in the educational market place: stories from the field of women in management. *Discourse*, 17(3), 337–52.

Blackmore, J. and Sachs, J. 2001. Women leaders in the restructured university, in *Gender and the Restructured University: Changing Management and Culture in Higher Education*, edited by A. Brooks and A. Mackinnon. Philadelphia: Open University Press, 45–66.

Bönisch-Brednich, B. 2002. *Keeping a Low Profile: An Oral History of German Immigration to New Zealand.* Wellington: Victoria University Press.

Bönisch-Brednich, B. 2008. Watching the Kiwis: New Zealander's rules of social interaction – an introduction. *Journal of New Zealand Studies, Special Issue: Watching the Kiwis*, 6–7, 3–15.
Bönisch-Brednich, B. 2011 (forthcoming). Strangers on campus: academic migrants, universities and the hiring of international excellence, in *Interkulturalität und Bildung*, edited by U. Kriebernegg, R. Maierhofer and H. Penz. LIT Verlag: Berlin et al.
Bourdieu, Pierre. 1990 [1980]. *The Logic of Practice*. Stanford: Stanford University Press.
Bourdieu, P. and Wacquant, L. 1992. *An Invitation to Reflexive Sociology*. Chicago: University of Chicago Press.
Bredella, L. 1991. Demonic Germans and naïve Americans: the dialectics between hetero- and auto-stereotypes, in *Mediating a Foreign Culture: The United States and Germany: Studies in Cultural Understanding*, edited by L. Bredella. Tübingen: Narr, 108–31.
Cefkin, M. (ed.) *Ethnography and the Corporate Encounter. Reflections on Research in and of Corporations*. New York and Oxford: Berghahn.
Ellis, C. 1998. I hate my voice: coming to terms with minor bodily stigmas. *The Sociological Quarterly*, 39, 517–37.
Ellis, C. 2009. *Revision: Autoethnographic Reflections on Life and Work*. Walnut Creek: Left Coast Press.
Friedmann, J. 2005. Place-making as project? Habitus and migration in transnational cities, in *Habitus: A Sense of Place*, edited by J. Hillier and E. Rooksby. 2nd edition. Aldershot, UK and Burlington, US: Ashgate, 315–33.
Giordano, C. 1996. Die Rolle von Mißverständnissen bei Prozessen der interkulturellen Kommunikation, in *Mit der Differenz leben. Europäische Ethnologie und interkulturelle Kommunikation* edited by K. Roth. Münster: Waxmann, 31–42.
Kolapo, F.J. (ed.) 2009. *Immigrant Academics and Cultural Challenges in a Global Environment*. Amherst and New York: Cambria Press.
Kreutzer, F. and Roth, S. 2006. Einleitung, in *Transnationale Karrieren: Biographien, Lebensführung und Mobilität*, edited by F. Kreutzer and S. Roth. Wiesbaden: VS Verlag für Sozialwissenschaften, 7–31.
Lippi-Green, R. 1997. *English with an Accent: Language, Ideology and Discrimination in the United States*. London: Routledge.
Rapport, N. 1995. Migrant selves and stereotypes: personal context in a postmodern world, in *Mapping the Subject: Geographies of Cultural Transformation*, edited by S. Pile and N. Thrift. London: Routledge, 267–82.
Rapport, N. and Dawson, A. (eds) 1998. *Migrants of Identity: Perceptions of Home in a World of Movement.* Oxford and New York: Berg.
Richardson, J. and McKenna, S. 2002. International experience and academic careers: what do academics have to say? *Career Development International*, 7(2), 67–78.

Saha, L.J. and Atkinson C. 1978. Insiders and outsiders: migrant academics in an Australian university. *International Journal of Comparative Sociology*, XIX (3–4), 203–18.

Scheibelhofer, E. 2003. *Mobilitätsperspektiven junger Wissenschaftlerinnen. Eine qualitative Studie zur Situation österreichischer ForscherInnen in den USA*. Projekbericht Wien: Institut für höhere Studien.

Schütz, A. 1944. The stranger: an essay in social psychology. *American Journal of Sociology*, 49, 499–507.

Shore, C. and Wright, S. 1999. Neo-liberalism in British higher education. *The Journal of the Royal Anthropological Institute*, 5(4), 557–75.

Sparkes, A.C. 2007. Embodiment, academics, and the audit culture: a story seeking consideration. *Qualitative Research*, 7(4), 521–50.

Strathern, M. 2000. The tyranny of transparency. *British Educational Research Journal*, 26, 309–21.

Ting, T. and Watts, M. 2009. Narrative inquiry: defogging expatriate expectations, in *Immigrant Academics and Cultural Challenges in a Global Environment*, edited by F. Kolapo. Amherst, New York: Cambria Press, 73–106.

Wright, S. and Ørberg, J. 2008. Autonomy and control: Danish university reform in the context of modern governance. *Learning and Teaching*, 1(1), 27–57.

Epilogue: The Cosmopolitan Justice of a Direction Home

Nigel Rapport

How does it feel
To be on your own
With no direction home
Like a complete unknown
Like a rolling stone?

Bob Dylan: *Like a Rolling Stone*

To be in hell is to drift: to be in heaven is to steer.

George Bernard Shaw: *Man and Superman*

I do not mean to begin pretentiously. Between the celebrity wordsmithery of George Bernard Shaw and Bob Dylan and the brute materiality of 'local lives' in mundane struggle, there can seem a chasm. And yet there is much common ground. It is the human terrain of finding the words to relate one's experience: every human being recounts their experience, to his or her own self as well as to others; every human being makes sense of experience in ways that prescribe at the same time that they describe. Wordsmithery is 'democratic': it is of people universally. More technically put, 'culture' is a rhetorical application: here is the global use of languages, verbal and other, to provide common forms in whose terms experience might appear to be shared and known, habitual and proper.

What I would point up is how, in their words, Bob Dylan and George Bernard Shaw appear to indicate a common ethos. It feels homely, even heavenly, to be in command of the direction one's life is taking; it is in this way that one feels that one knows one's life, that one belongs to it. This ethos has a special resonance, I shall say, for the experiences of movement with which this volume is concerned: homeliness is having a direction to life that one can recount, that makes sense (as both description and prescription), and that one determines.

The keyword of the present volume is 'local'. It is applied in ways that are instrumental, self-conscious and political: we witness the symbolic construction of locality. The volume understands a 'locale' to be not a set of over-determining mechanical linkages – as purportedly pertained in a so-called traditional social structure – but a space of identification and belonging (Cohen 1985). The particular emphasis I wish to give in this epilogue is on becoming. The identity of locality, the making of a home locale, is something that comes into being. That coming-into-being would seem to me to be its most important aspect. Making a home locale is an act of human creativity, individual or collaborative in character: it is a manifestation of a human nature, of a human capacity. I would wish for an anthropology that gives celebratory testimony to that capacity and that creativity, and I would wish for an anthropology that works to accord the capacity to create a home locale, as an act of becoming, the status of a human right. Part of a 'cosmopolitan' project of anthropology would be to espouse such home-making as something every human being can and could achieve: part of his or her life-project. All the time recognising, too, that home-making also encompasses home-breaking: the human capacity and the human right is to become, and no present structuration need accord with an individual's vision of how he or she would come to be at home; and hence no present structuration should determine or ascribe how an individual determines to become at home in his or her own life. Home is a directionality, one can say.

Let me show how this theme emerges, for me, from a volume of articles rich in evidence of individual lives.

Locality: The Politics of Place

This volume is intended to re-assess the phenomenon of 'locality' as a contemporary lived reality, the editors Brigitte Bönisch-Brednich and Catherine Trundle explain. Recent social-scientific attention being paid to transnationalism and mobility has entailed, it is felt, short shrift being given to the relevance of emplaced and localised identity. This is a mistake: the identity of the migrant continues to be constituted through land, bounded spaces and place-specific communities; the migrant continues to understand his or her experiences through spatial and material practices; and 'home spaces' continue to be vital to migrants and 'locals' – to 'hosts' and to 'guests' – alike. Albeit that a global flow of people, of ideas, goods, remittances and investments destabilise the seeming boundedness of home spaces, relations and members, locality does not wither away. Indeed, it gains a renewed salience as a marker of identity and of 'resistance' to the global and migrant threat: there is a complex interplay between global movement and discourses of localisation. Global and local enter into a discursive and experiential dialectic that is never static or unidirectional (Robertson 2010).

The particular focus of this volume is upon migrant responses to the discourse of localisation. The migrant, it seems, often endeavours to appropriate 'local' and place-specific symbols so as to claim belonging and articulate reconfigurations of the claims against them. The migrants themselves claim many kinds of emplaced ownership and entitlement in the host society. While re-validating the discourse of locality, however, this volume is far from offering a re-essentialisation of 'local'. The emphasis is on the processual, on localisation as cultural technique: here is how migrants utilise and engage diversely with ideas of locality. More specifically, a review of the substantive chapters reveals five main themes that exhibit themselves: they overlap and cross-cut.

The first theme concerns *the physicality of locality*. We hear that physical space is constitutive of locality in the Albanian coastal village of Dhërmi/Drimades (Nataša Gregorič Bon). Locality derives both from the villagers' sense of being 'rooted' to the place, land and property and also from their continuous movements and migrations across space. Likewise in the English Midlands town of Shepshed, local community is a manifestation of collective claims for place-based material resources and emplaced social networks of daily relationality (Karen O'Reilly). Community boundaries are drawn not foremost along lines of indigene versus immigrant but along real geographic and physical borders of the town and the material resources that 'locals' of all backgrounds fight to defend. The same physical, face-to-face contact makes St Joseph's Hostel a little pocket of Anglo-India in Melbourne's suburbs (Robyn Andrews). The rest home achieves this Anglo-Indian locality through a combination of décor, food, Christian rituals, accents and colloquialisms. The day-to-day practices and relations of the residents produce a 'local' space in an adopted environment of Melbourne, a physically homely retreat.

A second theme, however, concerns the way in which *locality is ever globally mediated*. The anthropologist knows this from personal experience: the modern university represents a field of tension, with local campus life firmly fitted into a global knowledge economy (Brigitte Bönisch-Brednich). The academic is a migrant who is compelled to play a dual role along her career path that merges local duties with global standards, opportunities and reputations. Similarly for our informants, such as middle class British émigrés to rural France, claims to locality should not be taken simply at face value for they exist within a circuitry of relationships in which people regularly engage and which extend outwards to uncertain extents (Michaela Benson). As these Britons mobilise their statements of local belonging, they continue to rely upon relationships with erstwhile compatriots elsewhere to confirm and reinforce their decision to migrate, to assess their position en route to a better way of life and to authenticate their post-migration lifestyles. Likewise in the Mallorcan village of Deia, what is now more than a hundred years of immigration and coexistence means that a division between foreigners and locals, outsiders and insiders, is not clear (Jacqueline Waldren). Now, under the guise of 'traditional village people' and a tightly formed community there is an acceptance of a fuller European or global identity: a rural idyll that is also a rendezvous for international jet-setters, *nouveaux riches* travellers and tourists.

A third theme concerns the way in which *the local sets the discursive terms of its global appropriation*. To be an American immigrant in rural New Zealand, then, is to routinely deploy the resources of identification that are locally available, since these afford the most moral purchase for making claims to local belonging (Catherine Trundle). However paradoxical, the American newcomers adopt those local anti-immigration discourses that circulate in a regional public sphere and they pursue a localness that affirms being rooted in the soil: virtuous custodian of the land. Likewise, for Bosnian Muslim immigrants in Melbourne, homeliness entails the ways in which common local discourses are taken up – reflected upon, negotiated with, challenged and critiqued – albeit that those discourses (an objectification of fundamentalist Islam) appear directly to challenge their fundamental sense of identity (Lejla Voloder). Becoming emplaced in local settings, configuring their belonging through daily encounters in streets, neighbourhoods and worksites, the immigrants construe a sceptical stance for themselves *vis-à-vis* the status of Australian multiculturalist inclusion. Even if being in a place is more necessity than choice, as with rural to urban migrants in the squatter settlement of La Ciénaga in Santo Domingo, residents invest in the production of locality – in the development of a relatively bounded material environment and the objectification of social relations within it – and do so by confronting the terms of their local exclusion from the domains controlled by the Dominican State (Erin Taylor). The result is that over the course of years, the squatter communities become permanent while remaining marginal and technically illegal, and while the reluctant settlers' aspirations continue to lie elsewhere.

A fourth theme, however, concerns the way in which *the local is a product of prior exterior (global) expectations*. This is the case, for instance, with those 'hippie' and 'drop out' migrants from the mainland United States to the Big Island of Hawai'i, and whom among the local inhabitants the immigrants choose to recognise as 'hosts' (Lucy Pickering). These migrants engage in processes of 'host-making', based on exterior assumptions about indigenous culture; these delegitimatise a local life in suburban bungalows beer-drinking and truck-driving in favour of a spiritually advanced autochthony embodying peaceful coexistance and a harmonious relationship with nature. This is similar to what we have heard of the British lifestyle migrants in the rural department of Lot in southwestern France (Michaela Benson): prior ideas about the locale inspired the decision to migrate as well as influenced expectations and experiences of life following migration. The local is already overwritten by the global.

A final and powerful theme concerns, then, *the indeterminate nature of 'locality'*: as with any other symbolic item in a cultural arsenal or fund of forms (including the 'global'), the 'local' is ambiguous and unstable. What is Deia as a 'local community' (Jacqueline Waldren)? It is both a tight physical enclave and also something without the geographic, social or artistic borders that might separate people from world events, social, artistic or environmental concerns, other villages, towns and countries, or family members and friends in faraway places. 'Community' is not a thing-in-itself, but a relational phenomenon that is variously articulated through actions, discourse and symbols. What is Lot as a 'locale' (Michaela Benson)? It is a concept variously utilised and defined by people as they negotiate their lives in a familiar and an unfamiliar setting. What is Dhërmi/Drimades as a 'locality' (Nataša Gregorič Bon)? It is an ambiguous form, continuously shifting as it is negotiated and managed by people pursuing practical, personal and social goals. 'Locality' can be defined as a process of becoming whose meaning derives from a continuous interplay between rootedness and mobility.

Moreover, such indeterminacy, I would like to say, is how it should be. Our cultural concepts should be tools that reflect our human becomings. Our concepts are the symbolic forms by which we hope to express our changing awareness of our lives, as individuals and as members of societies. And in that interplay between rootedness and mobility (between past and future) lies an expression of human freedom. We should be free to inhabit a set of cultural forms, a 'locality', and we should also be free to depart from this set and to exist in movement to other such sets or to none. We have the capacity, as individuals, to give meaning to cultural forms – they depend on our personal animation – and we should have the freedom to fulfil this capacity in ways that we personally see fit. This, at least, is the 'cosmopolitan' vision.

A Cosmopolitan Anthropology and the Human Terrain of the 'Glocal'

At the same time as he was describing the dialectical character of the 'global-human circumstance' – of integration alongside fragmentation, universalisation alongside particularisation, homogenisation alongside differentiation, essentialism alongside syncretism – Roland Robertson (1990) sought to prescribe a new human science that might properly address the concrete structuration of the world as a single ecumene. We need 'wide-ranging, empirically sensitive interpretations of the global-human condition as a whole', he wrote (1990:18). These will be interdisciplinary in character insofar as they borrow from natural science and the humanities as well as social science. Above all they will appreciate that the local and global are but poles of a continuum. We are witness at once to the ubiquitous global mediation of people's life-worlds and the ubiquitous localisation of global discourse. 'Global' and 'local' are mutually implicating principles: hence, more properly, the 'glocal'.

Others, too, have drawn attention to the circumstance of contemporary globalism as an apposite moment to re-assess the nature of its human-scientific accounting. For James Rosenau (1990), we cannot comprehend globalism within existing disciplinary premises. Sociology has tended to model social norms on an ideal-typical picture of the nation-state (reflecting the discipline's nineteenth-century routinisation in a nationalist era). It has tended to conceive of 'society' as a cohesive, ordered, bounded totality, comprising an integrated social system and a unified social space. This picture might always have represented a fiction, but even as such it is anachronistic: an obstacle to comprehending porous, fragmented and permeable social spaces, and multiple, overlapping and intersecting socio-spatial networks of power. Rosenau's conclusion is that we have the opportunity to offer an account of human affairs that deals in universals: a sociology that reaffirms the Enlightenment principle (scientific and ethical) of the singularity of the human. The opportunity is to aim 'at no less than the mobilization of Humanity itself as one self-conscious social agent' (Archer 1991: 146).

In anthropology, meanwhile, a significant aspect of the *Writing Culture* debate concerned what George Marcus and Michael Fischer (1986: vii) described as one of the discipline's major contemporary issues: how the contemporary 'inter-referencing' world is to be represented as an object for social thought. Is it appropriate to construe worlds of cultural difference, they elaborate (1986: 37–9), when socio-cultural units of study are spatially indistinct, when world systems penetrate and inform local worlds and vice-versa, and when the otherness that purportedly separated ethnographic subjects from home audiences is replaced by the pluralism of one social interdependency? A fetishisation of culture now becomes the domain less of analysis than of politics: the strategic essentialism of fundamentalists who would oppose both the liberal vision of culture as a lifestyle choice, a matter of taste, and the radical vision of a global political economy polarised between rich and poor, north and south. In short, 'culture' may no longer be an appropriate way anthropologically to signal critical difference: the moment is pregnant with other, necessary possibilities.

Finally, Ernest Gellner, while no friend of anthropology's post-modern mood of 'indulgent hermeneutic-subjectivist excess' (1992: 7) – namely *Writing Culture* – also described the need for anthropology to throw over the 'irresponsible affectations' of its recent cognitive relativism and moral relativism and to begin engaging with the modern world. 'We have no choice', he elaborated (1995: 8), but to 'work out the social options of our affluent and disenchanted condition':

> You cannot understand the human condition if you ignore or deny its total transformation by the success of the scientific revolution. The recognition of the inequality of cognitive claims in no way involves unequal treatment of people – quite the reverse. The accidental initiators of the scientific revolution have no monopoly on it, and are not necessary its best practitioners. Valid knowledge ignores and does not engender frontiers. One simply cannot understand our shared social condition unless one starts from the indisputable fact that genuine knowledge of nature is possible and has occurred, and has totally transformed the terms of reference in which human societies operate.

And just as trans-cultural knowledge represented 'the fact of our lives' – the starting point of any adequate anthropology – so working out a morality beyond culture was the pressing problem, however difficult to institutionalise (Gellner 1993: 54).

I mention these deliberations in anthropology and beyond because I wish to add to them. I subscribe to the explicit Enlightenment rationalism of the likes of Gellner and Rosenau: it is appropriate and necessary now to consider an anthropology that treats the singularity of the human, both in terms of a universal knowledge and a universal morality. I would call this a 'cosmopolitan anthropology'. It speaks to the dialectic between local and global modalities of the human condition, the 'glocal', and does so both in terms of universal human capacities and in terms of individual human rights.

The adjective 'cosmopolitan' appears in a number of the chapters of this book, as well as the editors' Introduction, by way of implying something mobile, diasporic or hybrid as opposed to something singularly place-based. I mean to confer a broader understanding, whose key is a reappraisal of the anthropological appreciation of culture. A 'global-human circumstance' is an opportunity to put culture in its proper place.

In a modern usage, the 'cosmopolitan' originates with the philosopher Immanuel Kant at the end of the eighteenth century. The term 'cosmo-politan' expressed a tension: to be human was at once to inhabit an individual embodiment, a particular spatiality and temporality (or '*polis*'), and to be a perfect instantiation of species-wide capabilities and liabilities, part of a finite human whole (or '*cosmos*'). Kant formulated his notion at the same time as he formulated the idea of 'anthropology' as a modern scientific discipline. Anthropology was the implementation of a cosmopolitan project: it was to engage in the study of particular human lives – lives conducted in particular spaces and locales – so as to derive understandings of human generalities. Traversing the spectrum between the human condition as it manifested itself in its individual particularities at one pole and in its species-wide generalities at the other gave on to both scientific insights and moral imperatives. The epistemology was dialectical. Anthropology came to know the human by zigzagging between individual and species: anthropology came to improve the human lot by regarding the particular individual life as an instantiation of the possible human life. Armed with a knowledge of the human whole, applications to improve the lot of particular lives could and should be made. Was a human potentiality being best served?

To describe culture as a rhetorical application – the point at which I began – and as a fund of symbolic forms, verbal and other, with which to express experience, is to assert an ontological position beyond culture: that of the agent who expresses and applies. It is here, too, that we have heard Gellner locating society, valid knowledge and global morality. For cosmopolitanism, crucially, the individual human being and the human species alike exist beyond culture: both possess an ontological reality – as things-in-themselves – which culture and its symbolic-rhetoric constructs do not. It is a human capacity to make cultures; it can be an individual human choice to inhabit a particular cultural world. But neither the human species nor the human individual, as things, are beholden to particular cultures: species and individual exist beyond their particular acts of making and their particular choices and tastes.

The cultural, in other words, is not a thing-in-itself: the cultural is a symbolic surface or clothing, a medium of expression, even a 'vehicle for a conception' (Geertz 1973: 208 n.19). But experience, expression and conception exist apart from it and remain responsible for providing the cultural with purpose, meaning and force: it is human usage, individual intentionality, which animates the cultural form and brings it to life in particular instances. One of the deliverances of globalism is a rendering of cultural surfaces more transparent: the distinctions between culture and humanity and between culture and individuality become more apparent. A human sameness is apparent beyond superficial cultural differences, and individual particularities are apparent beneath political rhetorics of cultural essentialism and homogeneity. *Cosmopolitanism entails insisting upon a recognition, which is both moral and rational, of the individual particularity and the human singularity that cultural forms might disguise.* It is a kind of emancipation. Cosmopolitanism would free the individual to explore liberally the space between what he or she is or would become, on the one hand, and how he or she would wish to seem (the forms of self-expression used, the collectivities and life-worlds joined) on the other. And cosmopolitanism would free humanity from formal, superficial (and invidious) differentiation into discrete parts (nations, ethnicities, religions, castes, classes). Martha Nussbaum's summary of the cosmopolitan programme is especially succinct:

> Whatever else we are bound by and pursue, we should recognize, at whatever personal or social cost, that each human being is human and counts as the moral equal of every other. … The accident of being born a Sri Lankan, or a Jew, or a female, or an African-American, or a poor person, is just that – an accident of birth. It is not and should not be taken as a determinant of moral worth. Human personhood, by which I mean the possession of practical reason and other basic moral capacities, is the source of our moral worth, and this worth is equal. … Make liberty of choice the benchmark of any just constitutional order, and refuse to compromise this principle in favour of any particular tradition or religion. (1996: 133, 136)

'Liberty of choice', to extend Nussbaum's final sentence, entails the individual's universal right to experience culture, to animate (create, adopt and adapt) its symbolic forms, in his or her own way and not necessarily in accordance with others' judgement of propriety, of tradition or of categorial ascription ('womanly', 'Muslim', 'respectable').

Such a perspective leads me to appreciate three further themes that emerge from the present volume of case studies. The first concerns *the constitution of a global society*. Ernest Gellner urged anthropology to consider the institutionalising of a global morality. With its expertise in the workings of social structures and the inventions of cultural traditions, anthropology might offer insight into the possibilities of kinds of global social-structuration which dealt justly with identity and difference. Indeed, the debate on global society, among anthropologists and sociologists, is a healthy one (for example, Smith 1990, Perlmutter 1991, Mathews 2000). Two features of such an ideal-typical society seem to me critical: first, *legal arrangements* whereby individuals enjoy the same conditions globally to nurture their own worldviews and to effect their own life projects; second, *norms of polite exchange* whereby individuals in interaction are ubiquitously afforded equal recognition of their distinct voices. A 'global constitution' is thus imagined to span an arc from legal formalities to polite or mannerly informalities. Bosnian Muslim immigrants in Australia (Lejla Voloder) should enjoy the global right, then, as individuals, to join and to leave, to form and to dissolve, to appreciate and to deprecate religious congregations. Rural migrants who squat in peri-urban areas of the Dominican Republic (Erin Taylor) have equal rights to gain access to local amenities (and duties to maintain these) so as to effect their aspirations of moving elsewhere. Newcomers and old-timers in scenic Mallorcan villages (Jacqueline Waldren) manage their interactions in terms of politenesses that overshadow differences of origin.

A second theme concerns *the sense of being at home in space*. Who is host and who is guest if the terrain of human beings is the glocal? I would like to say that in a cosmopolitan vision human beings adopt the roles of 'mutual guesthood', while it is the constitutional social structure of a just global society that hosts them. Whether individuals are old-timers or newcomers in a space, whether they are sedentarist or transient in their choice of life-project, one considers the constitutional arrangements (both informal and formal) to be the hosts of the individual practitioners who make use of them on a serial basis. The arrangements (both laws and politenesses) are kinds of contract. To be a farmer, an academic, a member of a religious community, a tourist, a child, comprises a routine kind of exchange between individuals and between individuals and the state. The ownership of land in rural New Zealand (Catherine Trundle), then, and the maintenance of a sustainable ecosystem in Hawai'i (Lucy Pickering), and the status and recognition of a new professor in a New Zealand university (Brigitte Bönisch-Brednich), are removed from the domain of personal or communitarian sovereignties – of local, indigenous or longstanding individuals and groups against newcomers or transients – and become instantiations of global, fair arrangements. There are contractual norms, instituted on the basis of universal equality and mediated by local conditions: any individual may act in accordance with them, for shorter or longer periods, while in pursuit of personal ends. The glocal is the home terrain of all and at any moment contractual norms host individual exchange and passage.

A third theme concerns *the right to one's unique becoming*, to a personal future. In Martha Nussbaum's cosmopolitan credo, a radical distinction is effected between the 'accident of birth' and the moral worth of human personhood. It should not matter where the individual happened to begin: as an exemplar of the human, that life possesses a universal value and is possessed of universal capacities for self-fulfilment. An individual life is a unique and absolute possession. The preciousness of that finite phenomenon is, so far as possible, to be vouchsafed from contingency: the life owes no dues to any one or any thing. The individual life is, after Kant's famous 'categorical imperative', a thing-in-itself and may not to be treated as means to any ends besides those it itself has construed. It is this which the constitution of a just global society endeavours to allow for: a space in which, so far as possible, each new individual may constitute their own lives, their own world-views and life-projects. A cosmopolitan constitution endeavours to afford each individual the space to come into his or her own – whatever newness and uniqueness that might entail.

The key here is futurity. The future may be regarded as the birthright of an individual life. Whatever the world-views and life-projects, the cultures and communities, and the religions and pastimes of those who had preceded him or her, whatever the desires and expectations of those who surround him or her in the present, it is the individual's right to become his or her own self as he or she sees fit, and to keep on so becoming throughout life. What is the future of lives, residents ponder, in coastal Albanian villages (Nataša Gregorič Bon), in the towns of the English Midlands (Karen O'Reilly), in the departments of rural France (Michaela Benson), and in an Anglo-Indian rest-home (Robyn Andrews)? One approaches these settings in terms of the individuals who would stay there or would pass through. One makes local conditions so far as possible embodiments of universal distributions of finite resources and universal knowledge concerning individual capabilities and liabilities: the individual is given the local space that he or she might be constitutionally assured anywhere to facilitate his or her becoming.

The cosmopolitan notions I have been spelling out are idealistic ones. But their normativity is well-served by the precedence of a volume such as this, bearing the stamp of experience, of authentic relations with individuals amid the current circumstances of their lives. A sense of 'cosmopolitan justice' might be said to manifest itself in the individual, always and anywhere embodying a directionality to his or her life that is personally achieved and not exteriorly ascribed. Whether the life is a celebrated one or is hermetic, to command its direction and to steer it as the individual would, is, as George Bernard Shaw and Bob Dylan urge, rightfully to become at home in a glocal space.

References

Archer, M. 1991. Sociology for one world: unity and diversity. *International Sociology*, 6(2), 131–47.

Cohen, A.P. 1985. *The Symbolic Construction of Community*. Chichester: Horwood.

Dylan, B. 1973. *Writings and Drawings*. London: Cape.

Geertz, C. 1973. *The Interpretation of Cultures*. New York: Basic.

Gellner, E. 1992. Anthropology and Europe. *Social Anthropology*, 1(1A), 1–7.

Gellner, E. 1993. *Postmodernism, Reason and Religion*. London: Routledge.

Gellner, E. 1995. Anything goes: the carnival of cheap relativism which threatens to swamp the coming *fin de millenaire*. *Times Literary Supplement*, 4811, 6–8.

Marcus, G. and Fischer, M. 1986. *Anthropology as Cultural Critique*. Chicago: University of Chicago Press.

Mathews, G. 2000. *Global Culture/Individual Identity*. London: Routledge.

Nussbaum, M. 1996. Patriotism and cosmopolitanism, in *Love of Country*, edited by J. Cohen. Boston: Beacon.

Perlmutter, H. 1991. On the rocky road to the first global civilization. *Human Relations*, 44(9), 897–1010.

Rapport, N. 1993. *Diverse World-Views in an English Village*. Edinburgh: Edinburgh University Press.

Rapport, N. 1994. *The Prose and the Passion: Anthropology, Literature and the Writing of E.M. Forster*. Manchester: Manchester University Press.

Rapport, N. 2003. *I Am Dynamite: An Alternative Anthropology of Power*. London: Routledge.

Rapport, N. 2011. *World and Anyone: The Cosmopolitan Project of Anthropology*. Oxford: Berghahn.

Rapport, N. and Dawson, A. (eds) 1998. *Migrants of Identity: Perceptions of Home in a World of Movement.* Oxford: Berg.

Robertson, R. 1990. Mapping the global condition: globalization as the central concept. *Theory, Culture and Society*, 7(2–3), 15–30.

Robertson, R. 2010. *Global Society*. Oxford: Blackwell.

Rosenau, J. 1990. *Turbulence in World Politics*. Brighton: Harvester Wheatsheaf.

Shaw, G.B. 1903. *Man and Superman*. Westminster: Constable.

Smith, A. 1990. Towards a Global Culture? *Theory, Culture and Society*, 7(2–3), 171–91.

Index

Page numbers preceded by 'fn' refer to a footnote, e.g. Africa fn5.

absentee owners 54–5
academic life 167–80
Adorno, Theodor 173
Africa fn5, 17
Agricultural Insurance Organisation 19
agriculture 52, 105
Ahmed, Sara 3, 7
Albania 9, 10, 17–28
alliances, emotional 38
Amélie, Dominican Republic 114–15
American dream 31–44
Amit, Vered 139
Anderson, Benedict 74, 139
Andrea, returnee 24, 25
Andrews, Robyn 9, 119–32, 185, 192
Anglicanism 128, 129, 155
Anglo-Indian Australasian Association of Victoria 123
Anglo-Indians 9, 119–31
Anthias, Floya 153
anthropology 5, 187–92
Appadurai, Arjun 2, 20, 73, 120, 121, 124, 130, 131
Arcadia 40–41, 44
Archer, Margaret F. 187
architecture 9, 53, 85, 104, 109, 110, 112, 159
Ardener, Edwin 97
art 86, 87, 186
Asad, Talal 116
Association France Grande-Bretagne (AFGB) 77–8
Atkinson, Claire 179
atmosphere 109, 124
Australia 119–31, 151–63, 186
Autochthony *see* locality
autoconstruction 104, 110
autonomy 21, 22, 26–7, 104–5

bach (holiday house) 36, 37
Bagnal, Gaynor 7–8
Balaguer, Joaquin 106
Bammer, Angelika 2
banks 142
bars *see* cafés/bars 137
Barth, Fredrik 71
Basch, Linda 72
Bashkow, Ira 5
Bauman, Zygmunt 2, 70, 101, 138
Bear, Laura 122
becoming 20, 183–92
being, ways of 73, 79
Belfast 6
belonging
  and being 73, 79
  and community 73–4, 139, 151
  and investment 139, 144, 146
  and land ownership 36, 42, 44, 61
  and language 169–70, 175–8
  and locality 71, 102, 104, 115, 131, 184–5
  myths 91–2, 94
  national 152, 154, 156–7, 162–3
  and place 3, 4, 6–8, 20, 24, 26, 27–8, 32, 72–7, 126
  and religion 156–62
  and social hierarchies 111–15
Bender, Barbara 90
Benson, Michaela 8–9, 67–83, 185–6, 192
Bharatiya Janata Party (BJP) 122
Blackmore, Jill 178
Bland, Martin 127

Boca Chica, Dominican Republic 103
Bönisch-Brednich, Brigitte 1–14, 126, 167–82, 185, 191
Borges, Jose Luis 87
Bosnians 151–63, 186, 190–1
boundaries 43, 49, 95, 135, 153, 185
  and place 5, 7, 88, 89, 95, 146–7, 171, 185
  social 27–8, 71–2, 90, 91, 95, 96–7, 186
Bourdieu, Pierre 8, 77, 80, 101, 108, 124, 171, 174
Brad, Hawai'i 52–3
Branson, Richard 85
Britain 7, 35, 70, 79, 122, 125
British migrants 9, 67–80, 177, 185, 186
British National Party (BNP) 145
Brooks, David 50
Buddhism 155
Buller, Henry 69
Burrell, Kathy 140
business systems 178

cafés/bars 91–2, 94, 113, 137
California 34
cannabis 49, 59
capitalism 96
Casey, Edward S. 7
Castada, Claudia 3, 7
caste system 119, 122, 125
Castells, Manuel 141
Castles, Stephen 36, 153
Catholicism 107, 108, 112, 115, 128–9
Certeau, Michel de 20
change 137–8, 141–2, 144, 148
Charnwood Borough Council 109, 143, 147, 167
Christianity 9, 18, 22, 27, 39, 122, 123, 155, 160, 162
  Anglicanism 128, 129, 155
  Catholicism 107, 108, 112, 115, 128–9
churches 86, 111, 112, 128, 144, 155, 159
citizenship 49, 151–63
Clifford, James 5, 18–20, 27
CODECI (Coordination for the Development of La Ciénaga) 108, 112
Cohen, Anthony 43, 71–2, 77, 139, 147, 183
collectivisation fn17, 18, 21, 22, 27, 28
colonialisation 32, 52, 57, 61
commerce 89
commitment 75, 79, 111, 115, 116, 136, 137, 138, 139, 146
  lack of 144, 145, 148
communication technology 20, 41, 57, 79, 95
communism 18, 19, 22, 23, 26–27
Communist Labour Union 23
community, local 67–8, 88, 107–8, 186
community, sense of 40–41, 96–7, 137, 138, 146
  and belonging 73–4, 139, 151
  lack of community spirit 73–4, 137–8, 140–2, 144, 145–6, 148, 167–8
  and place 5–6, 138–9, 141, 185
  see also identity
community boundaries 135, 185
community development 108
community involvement 41, 74, 115, 127, 148
conflict 23–6, 28, 54–5, 108–9, 114–15
connectedness 26, 32, 41–2, 88, 101
conservation 31, 42
constructivism 4
consumption 106–7, 111, 115
Cook, James 52
cooperation, community 107–8, 114
cooperatives, agricultural 22, 24
Cortàzar, Julio 87
cosmopolitanism 2, 6, 26, 49, 50, 51, 85, 86, 94, 188–92
criminalisation 112–13
cultural capital 25
cultural diversity 6
cultural identity 94–5, 156, 163
cultural understanding 172–5
culture 57, 103, 111
  academic 170–71, 172–5, 176–7, 178
  Anglo-Indian 125, 129, 131
  British 122, 125
  cosmopolitanism 50, 86, 94, 187–90
  multiculturalism 9–10, 96–7, 152, 153, 155, 156, 159–60, 163, 186

Muslim 155, 163
and place 72, 101, 139–40
shock 173, 178
and social boundaries 71–2, 90
Curtis, Jennifer 6, 38

Deia, Mallorca 85–97, 185, 186, 191
Delanty, Gerard 139
democracy 39, 105
depopulation 70, 74
Dhërmi/Drimades 17–28, 186, 192
diaspora 1, 120, 151
Dirlik, Arif 7
dirt 113
disability 136, 176
discipline 111, 151
disorder, social 105, 112–13
dispossession 106
Dominic, Hawai'i 49, 52–5, 59–60, 61
Dominican Republic 102, 103, 105, 191
Donald, Rod 36
dormitory towns 137–8, 144–5, 148
drainage 112
dress 156–61, 162, 169
drop outs 49–61, 186
'dwelling-in-travel' 19–23, 27, 28
Dylan, Bob 183, 192

East India Company 119
economic communities 95
economic crisis 95–6, 114, 140
Edith, Shepshed 138, 141, 144
education 93, 105, 107–8, 111, 128, 169
adult 142, 143
reason for migration 22, 23, 24
egalitarianism 36, 39, 177, 179
Ellen, Roy F. 57
Ellis, Carolyn 176
embeddedness 42, 78, 109
emotional connection 139–40, 144, 146
employment 22, 103, 104, 110, 122, 136–7, 140, 141–2, 148
England 10, 135–48
English language 122
Englund, Harri 4
Enlightenment 187, 188
environment issues 42, 53–4
Escobar, Arturo 101
ethnicity 18–19, 137, 139, 153, 155, 156–7, 159, 162
Ethnologia Europaea 6
Europe 17, 27
evangelists 39
exclusion 39, 71, 101, 105, 119, 122, 157, 159, 162, 186
existential mobility 49–50
expectations
of life following migration 49, 61, 70, 80, 173, 186
and lifestyle 85, 106–7
religious 152, 160, 163
exporting 103

fall of communism 23
families 33, 37, 74, 89, 109–10, 120
farms, family 36
Ferguson, James 26, 151
festivals 120, 142, 155, 161
Fischer, Michael 188
fishermen 54–5, 61
Fog Olwig, Karen see Olwig
food and diet 9, 120, 124, 125–7, 155, 167
Forbes magazine 85
Foreign Direct Investment (FDI) 34, fn36
Fortier, Anne-Marie 3, 7
France 8, 9, 67–80, 185, 186
Franco, General 96
freedom 138, 187, 190
religious 155, 158, 160, 161, 191
to travel 19, 20, 26, 103
Friedmann, John 171
friendships
with compatriots 78, 129
in home countries 41, 51, 77, 79
migrants with locals 70, 75, 80, 109
Frykman, Jonas 6
futurity 192

Gal, Susan 160
gardening 24, 76
Garendon estate 146–7
gated communities 31–44
Gedalof, Irene 157
Geertz, Clifford 189
Gellner, Ernest 188–90
gender issues 168, 174, 178

geometry, sacred 56
Geschiere, Peter 4, 17, 28
Giddens, Anthony 70
Giordano, Christian 174
Glassie, Henry 5
Glick Schiller, Nina 49, 72–3, 79–80
global perspective 88
globalisation
  global identity 96, 185
  and locality 1–4, 28, 72, 115, 120, 180, 185, 186
    'glocal' 2, 187–92
  and mobility 88, 103, 120, 138
  and neo-liberalism 172, 178–80
government advisors 52
gowns, academic 169
Graves, Robert 85
Great Mahele 52
Greeks 18–20, 23
Green Mountain 56
Gregorič Bon, Nataša 9–10, 17–30, 185, 192
Gregory, Steven 102–3, 110
guesthood 8–9, 49, 61, 94, 184, 191
Gupta, Akhil 26, 151

habitus 8, 124, 171
Hage, Ghassan 49
Haiti/Haitians 105, 106, 113–14
Hammarlin, Mia-Marie 6
Hannerz, Ulf 6
haoles 60, 61
Harvey, David 139
Hastrup, Kirsten 151
Hawai'i 8, 49–61, 186, 191
Hawaiians, Native 51–2, 57–60
heritage 3, 31, 44, 56
Herzfeld, Michael 80
hierarchies, social 102, 103, 111–15, 167
hijab 156–9, 161, 162
Himarë/Himara 17–28
Hindi 122
Hinduism 119, 120, 122, 155
hippies fn32, 49, 186
Hirsch, Eric 94
history
  local 25, 27, 74, 76, 89, 90
  national 156, 157
  shared 88, 96–7, 121, 136, 146, 147, 148
Hoey, Brian A. 41
Hoffnung-Garskof, Jesse 105, 111
Hoggart, Keith 69
holiday homes 37, 92
holiday resorts 85
home
  home making 126, 127, 130, 171–3, 183–92
  and place 2, 6, 7–8, 10, 51, 101–31
  space 160–1, 184–5
home ownership 22, 104, 111
Horlicks 127
horticulture 32, 90, 93
hosiery industry 137, 141–2
hosts 49, 60–61, 127, 186, 191
housing 33, 35–6, 52–3, 108–11, 112, 140, 142
housing, social 144
housing estates 145, 148
Hoxha, Enver 19
human rights 184, 188–92
hurricanes 106, 109

icons, national fn42
identity
  community 6, 108
  cultural identity 94–5, 156, 163
  and deterritorialisation 2, 71, 72, 101, 120
  and ethnicity 18–19, 153, 155, 156–7, 159, 162
  global 96, 185
  and locality 73–7, 136, 183–92
  loss 143, 144, 145, 147
  national 37, 50
  and opposition 34, 38, 43
  and place 3–4, 25, 32, 37–8, 44, 74, 184–5
  and religion 153, 154, 156–7, 158–9
  self-identity 7, 37–8, 73–7
  and space 2, 87–90, 95, 97, 184–5
  stereotyping 31, 37–44, 60, 75, 77, 104, 112, 114–15
  transnational 1, 31, 49, 103
  see also community
ill health 145

imagined community 74
immigration 3, 37–44
imports 111
inadequacy 173, 175–6
inclusion 116, 151, 155, 157–8, 162–3, 186
income fn104, 106, 111
India 122, 131
indigenous people 8, 49, 51, 52, 57–8, 60, 61, 101, 186, 191
individualism 115, 138, 148
individuality 138, 189, 192
industry 31, 32, 93, 94, 104, 137, 139, 141–2
inequalities 35–6, 39, 92, 115, 144–5
Ingold, Tim 76
inheritance 21–2
integration 17, 67–8, 73–4, 78, 79, 87, 90–91, 154, 171
investment, economic 27, fn35, 38–9, 43, 85
investment, emotional 139, 144, 146
Islam 152, 153–4, 155, 162, 163
islands 96
Ivakhiv, Adrian J. 56–7

Jackson, Michael 126
Jaliskari, Albania 25–6
Jewish Senior Citizens' Centre, California 129
Joesting, Edward 59
Jordens, Ann-Mari 123

Kant, Immanuel 189, 191
Kina coastline 32
Kirch, Patrick V. 58
Kofman, Eleonore 153
Kolkata (Calcutta) 122
Kosta, returnee 23–4, 25, 26

La Ciénaga 101–16, 186
La Clarin 113
La Residencia, hotel 85
Labour Party 23, 123
land ownership 17–29, 31–44, 51–5, 61, 101, 104, 185
land registration 25
landscapes 9, 76, 87, 88–90, 93–5
language 18, 52, 86, 94, 122, 158, 167–8, 170, 173, 175–8
latitude 19.5° 56
Law on Land 17, 18, 22–3, 25, 26, 28
Law on Restitution and Compensation of Property 23, 25, 26
Leicester 136, 137
Lemuria 59
Levitt, Peggy 49, 73, 79-80
Lewin, Erica 123
liberalism fn36, 105
lifestyle 69, 70, 79, 85, 96, 106–7
  migrants 8, fn32, 67, 69–70, 71, 73, 186
Liotta, Elena 7
living standards 36, 108
Lloyd-Webber, Andrew 85
localisation 1–4, 101, 102–3, 104, 109, 110, 116, 131, 184, 187
locality 20, 101, 141
  autochthony 4, 17, 28, 186
  and belonging 71, 102, 104, 115, 131, 184–5
  and globalisation 1–4, 28, 72, 115, 120, 180, 185, 186
    'glocal' 2, 187–92
  and housing 109–11, 115
  and identity 73–7, 136, 183–92
  and neighbourhood 120, 121, 130, 131
  and ownership 23–7
  and place 120, 124, 184–7
  threats from immigration 3, 37–8, 44, 85–6
localness 1, 8, 32–3, 42, 185–6
locals 142
  as hosts 8, 49–61
  and identity 88, 95
  relationships with migrants 68, 85–97, 157–8, 161, 167–8
    conflict 23–6, 28, 54–5, 56, 108–9, 114–15
    friendship 70, 75, 80, 109
  and tourism 23, 85–97, 103
Longhurst, Brian 7–8
Los Angeles Times Magazine 34
loss 140–42, 143, 144–5, 147, 148
Lot, France 67–80, 186, 192

Loughborough 137, 142, 143, 147
loyalty 24, 122, 151, 154

McAuliffe, Cameron 153
McGirr, Lisa 40–41
McSweeney, Gerry 35–6
Madison, Greg 51
Mallorca 8, 85–97
mangos 126
Marcus, George 188
marginalisation 102, 109, 110, 139, 153, 171
market regulation 36
Marlborough, New Zealand 35–6
Marquez, Gabriel Garcia 87
marriage 22, 89, 93
Marshall-Fratani, Ruth 4
Martínez-Vergne, Teresita 105
Massey, Doreen 7
material resources 7, 10, 135, 136, 140, 142, 143, 148, 185
materialism 92
meals 127
media 156–7, 159–60, 162
Melbourne 9, 119–31, 151–63, 186
memories 20, 37, 86, 88, 96, 135, 140, 146
men 75, 87, 91, 93
Menehune 55–6, 58–9
Meyer, Birgit 4
migrant routes 17–28
migrants
  academic 10, 167–80, 185
  American 8, 31–44, 185, 186
  Anglo-Indians 119–31
  in Britain 136, 137–8, 144–5, 146
  British 9, 67–80, 177, 185, 186
  drop outs 49–61, 186
  lifestyle 8, fn32, 67, 69–70, 71, 73, 186
  Muslim 152, 154, 186, 190–91
  part-time 41
  reluctant 101–16
  retirees 52–3, 61
  returnee 9, 10, 17–18, 22, 23–5, 27–8, 86
  rural to urban 22, 101–16, 139, 186, 191
  stereotyping 31, 37–44, 75, 77, 104, 112, 114–15
  urban to rural 6, 22, 31–44, 67–80, 85–97, 185
Migration Act 1958 (Australia) 123
Miller, Daniel 111, 140
Miller, Mark J. 36, 153
Mirsada, Muslim 155–61
missionaries 52
mobility 49–50, 95, 135–48
  and globalisation 88, 103, 120, 138
  lack of 2, 102–3, 136
  and loss of community 144–5, 148
  and rootedness 28, 50
  socio-economic 102, 103, 104, 115, 116
  and transnationalism 41, 184
modernisation 108
modernity 105, 106, 110
money 36, 39
morality 39, 44, 68, 105, 108, 113, 115, 188–91
Morton, Christopher 140
mosques 155, 156–7, 159
motorhomes 50
motorways 137, 138, 142, 145, 147, 148
multiculturalism 9–10, 96–7, 152, 153, 155, 156, 159–60, 163, 186
multinationalism 169
Muslims 152, 154, 163, 186, 190–91
Myerhoff, Barbara 129–30

national boundaries 49
National Heritage Fund 35
nation-state 17, 151, 187
Native Hawaiian Government Reorganization bill fn51
neighbourhood 90, 121, 130, 131, 152, 160
Nelson, New Zealand 31–4, 35–6, 40
Nelson Mail 37–8
neo-liberalism 139, 143, 172, 178–80
New Age pilgrims 56–7
New Zealand 8, 10, 31–44, 168, 185, 191
newsreader 156–7, 159
nostalgia 36, fn40–41, 107
Nussbaum, Martha 190, 191

Oak Creek, California 56
O'Hanlon, Michael 94
Olfosdotter Stensöta, Helena 6

Oliver, Caroline 74
Olwig, Karen Fog 80, 151
Ong, Aihwa 151, 153–4
opposition 34, 38, 43
O'Reilly, Karen 8, 10, 71, 74, 135–49, 185, 192
Ottomans 21, 22, 25
Overseas Investment Commission (OIC) 34–5, fn36
ownership *see* home ownership; land ownership; property ownership

Pan-Himarian conferences 26
patrigroups 22, 24, 26
Pebble Beach, California 34
performativity 178
philanthropy 39
Phillipson, Chris 139
Pickering, Lucy 8–9, 49–63, 186, 191
Pickett, Kate 144–5
pidgin 52
place
  and belonging 3, 4, 6–8, 20, 24, 26, 27–8, 32, 72–7, 126
  and boundaries 5, 7, 88, 89, 95, 146–7, 171, 185
  commitment 111, 136
  and community, sense of 5–6, 138–9, 141, 185
  and culture 72, 101, 139–40
  and home 2, 6, 7–8, 10, 51, 101–31
  and identity 3–4, 25, 32, 37–8, 44, 74, 184–5
  and locality 120, 124, 184–7
  and mobility 135–48
  politics of 5–8, 101–16, 184–7
  sense of 53–4, 96–7, 107
  and space 20, 38, 121, 127
places of worship 86, 111, 112, 128, 144, 155, 156–7, 159
Pocket Hawaiian Dictionary 58, 59
politics 5–8, 10, fn42, 85, 101–16, 184–7
Polynesians 58, 60
post offices 78, 137, 142
poverty 51, 103, 105, 112, 113–14, 139
power places 56–7
pre-marriage agreements 22
Press, newspaper 32–3
privilege 2, 113
progreso (progress) 103, 105, 109, 110, 111, 116
property development 31–6, 85–6, 90, 93, 146–7
property ownership 17–28, 33, 39, 69, 93, 94, 111, 185
property sales 33, 35–6, 95, 96, 142
public spaces 34, 36, 86, 89, 92, 106, 155, 160–61
  media 156–7, 159–60, 162
Puna, Hawai'i 53, 54
purity 113, 122

race and racism 106, 113, 115, 122–3, 145, 154
Rapport, Nigel 10, 71, 183–93
recreation 75–6, 120, 129–30
Redner, Harry 2
redundancy 69
refugee camps 101
relationships 28
  between compatriots in adopted country 8, 68, 77–8, 79, 157
  between immigrants of different race and religion 155–6
  between locals and immigrants 68, 85–97, 157–8, 161, 167–8
  conflict 23–6, 28, 54–5, 108–9, 114–15
  contact with family 9, 19–20, 23, 41, 51, 78–9, 109–10
    friendships 70, 75, 80, 109
  with compatriots in home country 51, 185
religion 9, 55–7, 111, 151–63
  see also Christianity; Hinduism; Muslims; places of worship
religious boundaries 153
religious diversity 154–6
religious expectations 152, 160, 163
religious freedom 155, 158, 160, 161, 191
religious sites 55–7
religious symbolism 156–7, 159–60
Rennie, Neil 57
research, academic 177
resistance 26–7, 52
respect 43, 92, 110, 114, 153, 177
respectability 109, 111, 112, 115, 190

responsibility 24–5, 27, 127
rest-homes 9, 119–31
retirees 52–3, 78–9
retirement 69
Robertson, Roland 2, 184, 187
Rodman, Margaret C. 89
romantic Orientalism 57, 58
rootedness 17–28, 32, 50, 185
Rosaldo, Renato 153, 158
Rosenau, James 187–8
Rothstein, Bo 6
Rousseau tinted 57, 60
rural idyll 6, 31–44, 67–80, 85–97, 185
rural to urban migration 22, 101–16, 139, 186, 191
Rusiñol, Santiago 85
RVers (recreational vehicle users) 50

Sa Tanca, culture centre 86
Sachs, Judyth 178
Saha, Lawrence J. 179
Sahlins, Marshall D. 58
Said, Edward 5
Samuel, Dominican Republic 107, 110
San Andrés, Dominican Republic 103
Santo Domingo, Dominican Republic 10, 101–16, 186
Savage, Michael 7–8
Schaeffer, Glenn 39
Schierenbeck, Isabell 6
schools 107–8, 112, 128
Schütz, Alfred 169–70
'second birth' 170–71, 176
secularity 153, 155, 156, 159, 162, 163
security 71, 104, 108
Sedin, Muslim 161
service industry 93, 94, 104
Shaw, George Bernard 183, 192
Sheller, Mimi 3, 7
Sheppard, Peter 89–90
Shepshed 136–8, 140–41, 143–4, 185, 192
shopping areas 86, 137, 140, 141, 142, 143, 144, 145, 146
shorelines 37
Smith, Bernard 57
snowbirds 41
social class 8, 103, 105–6, 139, 170–71
  middle classes 68, 69, 77, 79, 109, 110, 139
  working classes 105, 139, 143, 144
social cohesiveness 38, 41, 109–11
social differentiation 18, 27fn, 39, 177
social interaction 6, 40–41, 92, 127, 129–30, 178
social life 3, 10, 44, 49, 72, 91, 111, 140
society 187
solidarity 38, 39, 71, 96, 102
space
  and identity 2, 87–90, 95, 97, 184–5
  physical 120, 184–5
  and place 20, 38, 121, 127
  private 112, 113, 160–61, 163, 191
  public spaces 34, 36, 86, 89, 92, 106, 112–13, 155, 160–62
    media 156–7, 159–60, 162
  social 88, 94, 104, 157
Special Identity Cards for Aliens of Greek Descent 19, 23
spirituality 55–7, 59, 60, 128, 186
squatter settlements 101–16, 186
St Joseph's Hostel 119–31, 185, 192
St Mary's retirement village 123–4
stability 109, 111, 135, 137–8
stereotyping 31, 37–44, 60, 75, 77, 104, 112, 114–15
stories and myths 76, 91, 94, 96–7
Stratton, Jon 153
students 145, 174
Sunday Star Times 35–6
supermarkets 86, 95
Switzerland 35
Szanton Blanc, Cristina 72

Tasman, New Zealand 31–4, 40
Taylor, Erin B. 9–10, 101–17, 124, 186, 191
teaching 177
territorialism 7, 101
  deterritorialisation 2, 71, 72, 101, 120
terrorism 153–4
Thompson, Paul 139
Tilley, Christopher 90
Ting, Teresa 172, 175
toilets, public 142

Tomas, Tim 89–90
Torresan, Angela 80
tourism
  and land ownership 17, 18, 23–5, 26–7
  and locals 23, 85–97, 103
tradition 74, 94
transnationalism 1–4, 71–3, 111, 116, 131, 151
  and identity 1, 31, 49, 103
  and mobility 2, 41, 103, 107, 120, 184
Trask, Haunani-Kay 58
travel freedom 19, 20, 26, 103
Trent, Alan 32–3, 37, 38–9, 42
trespassing 54–5
Trujillo, Rafael Leonidas 106
Trundle, Catherine 1–14, 31–48, 185, 191
Turner, Victor 129–30
trust 145

United States 35, fn40–41, 52
universities 167, 171, 178, 185, 191
urban living 101–16, 135–48
urban to rural migration 6, 22, 31–44, 67–80, 85–97, 185

Verdery, Kathrin 25, 27
Vichy regime 70
victimhood 42, 135
Victoria University of Wellington 167–80
Victorian Anglo-Indian Association 123
village council 85
virtual communities 139, 148
Voloder, Lejla 9, 151–65, 186, 190
volunteer centres 142
vulnerability 31, 42

Wacquant, Loïc J.D. 124
Waldren, Jacqueline 8–9, 85–98, 185–6, 191
Walter, Richard 89–90
water supplies 53–4, 93
Watts, Michael F. 172, 175
wealth 39, 93, 105, 106, 110, 113
Welch, Christina 57
welfare, state-funded 51, 95
Wellington, New Zealand 167–80
Wilkinson, Richard 144–45
Wilson, Peter 112
women 75, 87, 93, 127, 174, 178
wool industry 137
writing, academic 176

Yiddish communities 129–30
young people 85, 88, 89, 91, 93, 94, 95, 113, 148
Yuval-Davis, Nira 153

For Product Safety Concerns and Information please contact our EU
representative GPSR@taylorandfrancis.com
Taylor & Francis Verlag GmbH, Kaufingerstraße 24, 80331 München, Germany

www.ingramcontent.com/pod-product-compliance
Ingram Content Group UK Ltd.
Pitfield, Milton Keynes, MK11 3LW, UK
UKHW021720240926
475702UK00029BA/1707
* 9 7 8 1 1 3 8 2 5 0 9 6 3 *